The publisher gratefully acknowledges the generous contribution to this book provided by the Stephen Bechtel Fund.

Climate Change in California

Climate Change in California

Risk and Response

Fredrich Kahrl
and
David Roland-Holst

UNIVERSITY OF CALIFORNIA PRESS

Berkeley   Los Angeles   London

University of California Press, one of the most distin-
guished university presses in the United States, enriches
lives around the world by advancing scholarship in the
humanities, social sciences, and natural sciences. Its
activities are supported by the UC Press Foundation and
by philanthropic contributions from individuals and institutions. For more information, visit www.ucpress.edu.

University of California Press
Berkeley and Los Angeles, California

University of California Press, Ltd.
London, England

Library of Congress Cataloging-in-Publication Data

Kahrl, Fredrich
    Climate change in California : risk and response /
Fredrich Kahrl and David Roland-Holst.
        p.   cm.
    Includes index.
    ISBN 978-0-520-27181-4 (pbk., alk. paper)
    1. Climatic changes—California.   2. Climatic changes—
Economic aspects—California.   3. Climatic changes—
Environmental aspects—California.   I. Roland-Holst,
David W.   II. Title.
QC903.2.U6K35   2012
363.738'7409794—dc23   2012003935

Manufactured in the United States of America

21   20   19   18   17   16   15   14   13   12
10   9   8   7   6   5   4   3   2   1

In keeping with its commitment to support environmen-
tally responsible and sustainable printing practices, UC
Press has printed this book on 50# Enterprise, a 30% post
consumer waste, recycled, de- inked fiber and processed
chlorine free. It is acid-free, and meets all ANSI/NISO
(Z 39.48) requirements.

CONTENTS

PREFACE

It is difficult to overstate the benefits humanity has enjoyed from domesticating carbon fuels. Burning wood provided essential safety, food, habitat, and technology services for most of our history. Then, about two centuries ago, we began an innovation process that translated the latent energy in fossil fuels into a miraculous array of productive uses. This Promethean gift of the sun, captive in prehistoric biomass until we began our industrial revolution, has conferred on many living standards beyond what our ancestors could have imagined. The scope of this prosperity remains limited, but it is clear that our energy-driven technology revolution is irreversible and indeed has accelerated in recent decades. Now, however, we are awakening to an unintended and unwelcome side effect of our carbon fuel renaissance, the climate-altering influence of greenhouse gas emissions.

Changing climate processes will ultimately change our lives, and a better understanding of these impacts can help public and private actors adapt more effectively and perhaps even change the behavior that brought this global environmental threat upon us. To support broader awareness of the climate challenge, this book reviews the most recent economic evidence on climate risk for California, the world's eighth-largest economy and one of its most diverse. In these pages, readers can see examples of how pervasive the impacts of climate change will be; it will affect every form of economic activity, our natural resource security, and the livelihoods we all derive from these. For example, climate change will both directly and indirectly threaten trillions of dollars worth of public and private assets over the next two generations, primarily through escalating fire, sea level, and storm risk. At the same time, natural resources such as water, forests, and montane and coastal recreational assets will be seriously threatened. Although California is unique, the challenges it faces will be mirrored in diverse ways everywhere, and we hope the insights presented here will raise awareness, further stimulating climate change research and more proactive policy dialogue across the globe.

All these changes, and the new risks they pose, will require changes in human behavior at the individual, social, and political levels. We may feel individually powerless to change the path of global emissions that is changing our climate, but each of us has a responsibility to protect ourselves from its consequences. As we learned from the San Francisco earthquake of 1906, systemic environmental risk requires us to adapt, changing behavior and investing to protect ourselves from forces we may not control. The sooner society awakens to the reality of climate change risk, the more effectively we can begin adaptation to limit costs to present and future generations. Indeed, we argue in this book that climate risk presents a new opportunity for knowledge-intensive innovation, a post-industrial revolution that supports our aspirations to prosperity in a lower-carbon, climate-altered future in which we can continue economic progress without endangering our environment and ourselves.

ACKNOWLEDGMENTS

For financially supporting the research behind this book, we thank the foundation Next 10. For outstanding research assistance, our thanks are due to Sam Beckerman, Drew Behnke, John Chen, Henry Ching, Elliot Deal, Sam Heft-Neal, Adrian Li, Xian Ming Li, Katarina Makmuri, Joanna Nishimura, Cristy Sanada, Lawrence Shing, Sahana Swaminathan, and Rainah Watson. Wendy Tao contributed both text and insight to the chapter on transportation.

Noel Perry, Morrow Cater, Sarah Henry, John Andrew, Ralph Cavanaugh, Guido Franco, and Adam Rose offered helpful comments. We have also benefited from the research insights of many colleagues, all dedicated to advancing our understanding of climate change and the challenges it presents, and from the work of PIER and other California initiatives, which have generously supported much of the research we cite here. We have attempted to cite original research contributors in all cases when their ideas and findings are mentioned in this book. Opinions expressed here remain those of the authors, as do any expository and interpretive errors, and should not be attributed to the institutions with which they are affiliated.

Introduction

The Golden State: so much of the ethos of California, in the minds of those who live here and around the world, suggests opportunity, prosperity, and natural beauty that it is difficult to imagine a future of hardship and adversity. Even in the worst time of modern American economic history, the dust bowl era, California was a beacon for countless migrants seeking a better future. Today we a face a new global threat from climate change, one that portends dramatic adjustments in our natural world and the way we live in it. How will California and its economy adapt, and at what cost? This book reviews the latest evidence on how we can expect climate change to challenge us, putting our livelihoods and assets at risk, and how we can change our behavior and institutions to achieve more sustainable prosperity.

The Earth's average surface temperature has begun an upward trend that is largely irreversible over the next century, regardless of efforts to reduce greenhouse gas emissions. Whether this trend turns out to be moderate or extreme will depend on policy, but in any case extensive environmental change, with attendant economic adjustments, can and should be anticipated. Some areas of the world, especially those with poor majorities living close to sea level, may be catastrophically affected. California, for the opposite reasons, will not be. At the state level, climate change need not be thought of as an asteroid strike, and addressing it can appropriately be compared to the challenge of steering a supertanker to avoid a distant collision. Individual economic interests in the state may experience dramatic climate impacts, but the state as a whole has the means to avert large-scale adverse consequences. The extent of the state's success in this will depend on foresight, policy determination, and private agency.

This book supports proactive climate response with a multisector assessment of California climate risk and response. It includes a comprehensive review of the available evidence on future climate processes and potential damages, economic assessment of this damage, and a review of options for managing climate risk and securing the basis for a sustainable future. This chapter begins by summarizing our main findings, then reviews the salient opportunities and challenges that climate change presents to the people, enterprises, and public institutions of California.

MAIN FINDINGS

From the most general perspective, our review of existing research on climate risk suggests three findings.

First, at the aggregate level, California has the economic capacity to adapt in response to foreseen climate risk, but doing so effectively will require better information, determined policy making, and extensive institutional adaptation. Prevailing estimates indicate that if no action is taken, climate risk damages would total tens of billions of dollars per year in direct costs, lead to even higher indirect costs, and expose trillions of dollars of assets to collateral damage risk. Climate response, on the other hand, can be executed for a fraction of these costs through the strategic deployment of existing resources for infrastructure renewal and replacement as well as significant private investments, which can stimulate both employment and productivity. Damages will take two primary forms: effects of long-term shifts (e.g., average temperature increase) and effects from a higher frequency of extreme events (e.g., heat waves). The former will be gradual and easier to adapt to; the latter, which we may already be experiencing, will escalate public and private costs abruptly and (in the short run) unpredictably.

Second, there will be some very significant adjustment challenges at the sector level, requiring as much foresight and policy discipline as the state can mobilize. In this sense, the political challenges may be much greater than the economic ones. The state's adaptation capacity depends upon flexibility, but divergence between public and private interests can inhibit this adaptability, making long-term risks to society higher than necessary.

Third, despite the impressive scope and quality of existing climate research to date, the degree of uncertainty regarding many important adjustment challenges remains unacceptably high, and information on nearer-term adaptation options is completely inadequate. This uncertainly is costly, increasing the risk and potential magnitude of mistakes, including the deferral of necessary adaptation decisions. Further improving understanding of climate effects may itself be costly and difficult, but policy makers need much better information regarding climate risk and response options.

California can respond to climate risk by developing effective response strategies, including defense (against adverse impacts like a rising sea level) and adaptation (shifting to more sustainable growth patterns). A real commitment to this could begin immediately by establishing and extending the capacity for technical assessment and policy analysis, followed by timely and sustained policy activism. California's historic Global Warming Solutions Act (AB 32) initiative is a positive model for this, but it is only a beginning.[1] The scope of long-term climate issues is much wider than the greenhouse gas mitigation agenda, and addressing climate change could sustain a broader and longer-term spectrum of growth-oriented investment and innovation.

OPPORTUNITIES

Although it is forbidding and uncertain, climate change will definitely present opportunities as well. Adaptation is in its essence a learning process, and thus Californians today and in the future can benefit from climate change by learning more sustainable patterns and practices of living and working. This may at first sound like social ideology, but it is very pragmatic. Humanity has overcome all the critical challenges presented to it over the ages—most of them natural but some self-inflicted—and our societies have emerged stronger and generally more prosperous. Worldwide wars in the last century led to new institutions of multilateral reconciliation and new global economic systems for more effectively sharing resources and technology. The threat of large-scale starvation from population growth has accelerated agricultural innovation since the 1970s in ways that have significantly improved global health and livelihoods. In the same way, the challenge of climate adversity can arouse societies to constructive change, innovation, and proactive adaptation that will make the future better than the past or present.

As it did with implementation of the New Deal and the creation of the Department of Homeland Security, the government can turn adversity into a growth opportunity with policy leadership. Proactive measures such as new and renewed public expenditures on infrastructure (e.g., the new San Francisco–Oakland Bay Bridge) can stimulate local job creation and, by the first-mover example of public commitment, complementary private investments. Investment incentives and promotion of technologies for adaptation, such as better home insulation, more efficient air conditioners, and technologies as yet undiscovered, should also be included in public policy measures. Adaptation can foster innovation and knowledge-intensive solutions to the challenges we shall all face. As it did in information technology and biotech, California can take leadership in this emerging industry, solving the state's problems while developing globally competitive technologies.

CHALLENGES

Climate change is, of course, the primary challenge we consider in this volume, but understanding its impact and how to mitigate that reveals many other challenges. Foremost among these are uncertainty and institutional capacity. We discuss them conceptually in the next chapter, but the two have many very practical dimensions.

Any effective response to something as important as climate change must be based on credible and timely information, including evidence about the costs of doing nothing and about the efficacy of as many alternative responses as possible. The first priority of most of the analyses we review is to establish the real risk and cost of inaction, or what will happen if we passively stand by and watch climate change take its course. The available evidence shows that California will pay a very high price for inaction, but this evidence, which requires predicting future events, is essentially uncertain and therefore only partially effective in motivating action. Research on adaptation options is even less advanced, and, given the broad spectrum of adaptation options, evidence here is even more uncertain. Some have used this uncertainty to completely defer decisions about responding to climate change, but this is not a responsible approach. We know from many other aspects of our modern lives that economic risk can be managed at an acceptable cost. Indeed, a multitrillion-dollar global insurance industry exists for this reason, and billions have likewise been invested for earthquake engineering in California. The appropriate way to manage climate risk is to invest in better information and identification of hedging—or, in this case, adaptation— options. This reasoning yields one of our most important conclusions: much more research is needed to support timely and effective climate adaptation.

In the adaptation context, institutional challenges may be even greater than those posed by systemic uncertainty. We think about climate change first in terms of physical processes, and then impacts, and thus it is natural to visualize adaptation in physical terms, such as levees and seawalls. In reality, effective climate adaptation will require a combination of hard and soft (institutional) infrastructure. Even when an engineering solution exists to mitigate an adverse impact that is occurring or we believe is imminent, our existing institutions may not be equal to the task of responding. This means that credible perception of climate change may be necessary to arouse our response, but that response will be insufficient without institutional change. We present many such examples below, but the general truth is that today's system of politics and public administration in California is not well suited to adaptive management; successful adaptation will require institutions that promote a more rational political approach to natural risk.

Climate change may be global, but its impacts will primarily be seen as local. Moreover, only a small fraction of policy and social institutions can influence global greenhouse gas emissions, but every community and household has a responsibility to protect itself from climate change's adverse effects. For these reasons, the adaptation agenda belongs to everyone, and every social institution has a stake in it. Indeed, local institutions are often in a better position to respond to local challenges because of congruent stakeholder interests, better information, and more immediacy.

The localization of adaptation also makes sense from a jurisdictional perspective. Since climate change is a problem that subnational and especially local governments cannot address directly, state and local governments do not directly control the likelihood of climate impacts. However, they have far more autonomy over adaptation (e.g., land use planning, water resources management, and emergency response preparedness) and should be directly motivated by evidence on climate damages.

At the state level, policy makers face several institutional challenges. The first is determining their place in the adaptation policy hierarchy, deciding how much authority and responsibility to delegate to local governments, and determining how to share authority with the federal government. Of course, existing arrangements are supported by a long history of precedence, but these arrangements may not be well suited to the financial and management needs for adaptation.

Second, the institutional dialogue and evolution that is needed for adaptation will necessarily include the private sector, but this presents both challenge and opportunity. Because climate impacts are so pervasively intertwined with resources and property, little policy adaptation will be possible without historic changes in public-private resource and risk-management partnerships. Because these issues project economic risk and substantial new public financial commitments across one of the world's most diverse economies, distributional issues will also complicate the dialogue. This said, more inclusive, collaborative institutional frameworks can enlist the enormous financial and innovation potential of private agencies to help California address this threat, and this opportunity should not be missed.

There is a very active debate in the academic literature about how much of the adaptation burden should be borne by the private sector. Some believe private agency can better anticipate and respond to the challenges of climate change, with market efficiency arguments that suggest decentralizing decision making to rational individuals will minimize aggregate impacts. Although these market efficiency narratives may be soothing to some, they give no comfort to those who, after reading this book, recognize that adaptation presents many trade-offs and its costs and benefits fall very unequally across society. Because of this, the public interest will have to play a superior role in deciding the course of climate adaptation.

Third, integrating adaptation into state planning processes will require more systematic approaches to cost-benefit analysis and decision making. In the case of infrastructure, for example, the state can take advantage of experience and large recurrent maintenance budgets, but it needs clear guidance about how to formulate and time priorities for adaptation investments. Currently, most climate research is not about policy or economics and in any case extends over time horizons too long to inform these decisions. Much of the research on adaptation has actually been about evidence on physical climate impacts, attempting to build the case for mitigation rather than developing the frameworks and tools needed to make difficult political decisions about how, when, and where to allocate public goods and services. Capacity for public adaptation strategy, when it does emerge, will have to be supported by more relevant, shorter-term risk assessment.

WHAT IS AT STAKE

All this discussion about risk and strategy begins to sound like gambling, and of course our responses to climate change are a form of gambling. If we choose inaction, then the odds are out of our control, but if we adapt we can shift the odds in our favor. The amount of effort and resources we are willing to devote to adaptation depends on the stakes involved, and most of the rest of this book is dedicated to evidence that climate change is a very high-stakes game indeed.

This book is about the challenges and opportunities that climate change poses for California. With a gross state product of $1.9 trillion in 2010, California is the largest state economy in the United States, accounting for 13 percent of U.S. gross domestic product. California's economy is very diverse, driven by the manufacturing, trade, information, and services sectors. Primary industries—those that are directly dependent on natural resources—play only a minor role (fig. 1).

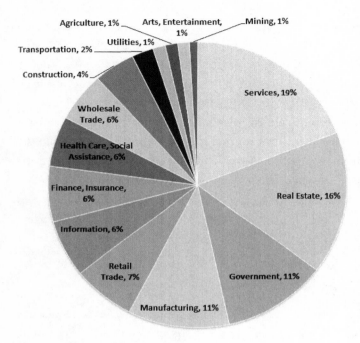

Figure 1. Shares of California real gross state product in 2005. Agriculture here includes agriculture, forestry, and fishing. Source: Data are from the Bureau of Economic Analysis (BEA) website, "Regional Data," http://bea.gov/iTable/iTable.cfm?reqid=70 &step=1&isuri=1&acrdn=3.

Of the sectors likely to be affected by climate change, only real estate and health are currently a major part of the California economy, although climate change may extend its influence by driving up resource prices. In this study we base our assessment on real activity across seven sectors of the state economy:

1. Agriculture, forestry, and fishing
2. Water
3. Energy

4. Transportation
5. Tourism and recreation
6. Real estate and insurance
7. Public health

For each of these sectors, we dedicate a chapter in which we review evidence on the potential physical impacts of climate, assess their economic costs and the distribution of these costs, and discuss adaptation opportunities and challenges.

We chose the seven activities for their climate relevance, either because they are resource intensive or climate vulnerable. They are also of interest because of their diversity in terms of the goods, services, employment, and income they provide the state, and also in the ways they are privately managed and publicly regulated. For example, energy (electricity) and water have highly regulated, cost-of-service pricing and generally charge customers at or below average cost. Agriculture gets state support for water costs and receives federal subsidies to compensate farmers for major weather disasters, but in general insurance companies do not insure crops. Real estate and public health are both backed by the insurance industry, which means that impacts and risk pooling are based on averaging in insurance premiums. Transportation infrastructure is generally not paid for by users but through taxes, a more extreme abstraction from the principal of beneficiary pricing.

Each sector's particular material characteristics, management, and regulatory history will suggest the path of adaptation most appropriate to it. In regulated sectors, average cost pricing can help to smooth out cost fluctuations due to climate change and adaptation needs. On the other hand, none of the regulated sectors is particularly flexible because of high concentration, long investment cycles, and regulatory processes themselves. This trade-off between closer public sector engagement and flexibility presents an important challenge to adaptation policy. Clearly the public interest needs committed representation in some of these strategic sectors, particularly against large long-term risks such as climate damage. Somehow, however, public-private partnership must evolve in ways that are more congenial to adaptability and innovation.

CHAPTER ONE

Economic Perspectives on Climate Adaptation

Scientific evidence has established a connection between economic behavior and climate change, primarily through the use of carbon-based energy sources in pursuit of higher material living standards. Over the last two centuries, domestication of carbon fuels for industrialization has conferred on a large share of humanity living standards that would have been beyond the imagining of their forebearers. Recently, however, we have awakened to the fact that this Promethean gift is changing the natural world in fundamental and adverse ways. This unintended negative externality has given rise to what is often called the "mitigation agenda," a local, national, and global policy dialogue about how to limit greenhouse gas emissions and other anthropogenic contributions to climate change.

Conversely, climate change has begun and will continue to present to the economy and all its participants a broad spectrum of challenges as well as opportunities. The general response to these challenges is referred to as the "adaptation agenda." Although both mitigation and adaptation are essential environmental issues and unified in the context of climate change, it is important to recognize their differences. Mitigation can, and probably should, improve environmental conditions and slow the progress of climate change. Mitigation activities, which are being discussed, negotiated, and promoted at all levels of society, represent a complex agenda of social responsibility and environmental citizenship from the global to the individual level. Because mitigation is causative (i.e., behavior leads to an effect) and this externality is global, cooperation is essential to progress.

The adaptation agenda is fundamentally different because it is responsive (i.e., effects induce behavior) and more localized. Greenhouse gas emissions are a local activity that affects the global environment, but climate change is a global process that affects localities. For this reason, individuals, communities, and even states can debate their role in the mitigation agenda, but everyone has a responsibility to protect themselves from the adverse effects of climate change. Regardless of what we do about emissions, climate change has begun and will continue for generations. The extent of change will depend on mitigation decisions, but adaptation is inevitable.

Effective adaptation requires an understanding of the challenges and opportunities alluded to above. In this chapter, we review the basic economics of adaptation, including its costs and potential benefits. An economist's perspective offers only part of the insight society needs to advance the adaptation agenda, but the concepts presented here can elucidate how a changing climate will create economic risks and rewards. As they always have, these two incentives can be expected to animate human behavior and move society along the path toward economic sustainability. Public agency is also discussed here because it will be essential to facilitate timely, cost-effective, and inclusive climate adaptation. Like private actors, policy makers need incentives to animate and change their behavior, and the concepts we cover below also recognize their essential role and the risks and rewards they face.

In this overview of the basic economics of climate adaptation, we present conceptual principles that provide a foundation for the chapters that follow. Because climate adaptation is localized, some of the lessons from California will be more generally relevant than others. The general conceptual principles, however, should be relevant anywhere. We discuss five topics of special relevance to climate adaptation: uncertainty, costs, timing, fairness, incentives, and institutions.

UNCERTAINTY

An essential consideration for all adaptation decisions is uncertainty and its economic avatar, risk. As we all know from the weather report, uncertainty is endemic to our understanding of climate processes, and climate-economy linkages only compound that uncertainty. We can never expunge uncertainty completely, but individual and social responses to it can be more constructive if better information about events and consequences is available. The importance of this uncertainty for economics is that it complicates decision making and can affect behavior in pervasive and often socially undesirable ways. Throughout this book we argue that one of the main barriers to adaptation is inadequate information available to private and public actors. Generally speaking, the available evidence is not relevant or authoritative enough to support timely and decisive policy.

Uncertainty in the context of adaptation takes six forms:

1. Climate processes
2. Baseline economic resource availability and activity
3. Impacts of climate change on economic resources and activities
4. Institutional constraints and capacity to respond to climate change
5. Technological change
6. Responses to adaptation measures

This book focuses on developing evidence for the third and fourth areas, economic impacts and institutional response, because we believe a stronger analytical foundation, a candid discussion of risks and trade-offs, and greater consideration of the current and future effectiveness of institutions are most needed to advance proactive, sound decision making. At the present time, the vast majority of information about climate change, scientific evidence presented in terms of physical systems, is related to the first category. Financial markets, the insurance industry, and indeed most of the rest of the economy are not managed by scientists, and economic agents need to see climate costs and benefits in material terms if they are to commit resources for adaptation. Government agencies need to understand the costs and benefits of different adaptation options if they are to make decisions that put public resources to their best use.

More assessment research is also needed to improve the quality and authority of evidence on climate change. This will help overcome an important barrier to effective policy response, disagreement among stakeholders about the facts of climate change and its impacts. Part of our motivation for writing this book was a desire to translate scientific evidence into implications that society can interpret and act upon, namely real and potential economic impacts. Information in this form can support responsible risk assessment and strategic responses to uncertainty. It will also improve the preconditions for cooperative solutions and public agency.

COSTS

Two kinds of costs are relevant to economic decisions about climate change: the costs of climate damage and the costs of adaptation. This book is mainly about the former, but better information on what it costs to adapt will also be needed to move forward.

Because climate has such a complex and pervasive relationship with the natural world, with resources, and therefore with the economy, there is a range of approaches to considering the economic impacts of climate change. In some cases, scientific information translates relatively directly into economic variables. For example, temperature-induced changes in crop yields can be converted to changes in expected harvest, and shortfalls can be assessed across a range of expected market prices. In cases where uncertainty is less specific, we need a more indirect approach to risk valuation.

This simplicity can be contrasted with the complexity of more general climate risks, like fire risk to property. Unlike in the case of a farmer planting a specific crop, in this case there is no model of fire incidence, severity, or timing to credibly inform individuals or communities about expected costs that are specific to their circumstances. In contexts such as this, the best we can do is give estimates of total assets at risk and actuarial averages for fire frequency and severity. Given that the past may not be a reliable guide to the future, the insurance industry will need new approaches to risk assessment that provide the public with a better sense of the changing cost of risk.

The problem with this approach reveals another obstacle to effective adaptation decisions: the average versus the marginal cost of risk. To get fire safety incentives right, for instance, property owners should pay for protection in proportion to their individual (marginal) contribution to overall fire risk. Fire defense based on risk pooling and average cost calculations is generally inefficient because of disparities in initial risk and ultimate damages. In a typical fire, some proportion of assets is catastrophically damaged while others are unscathed, with the former usually including a disproportionate number of higher-risk properties. Risk pooling thus effects a financial transfer between these groups. Risk-based pricing, where insurance premiums better reflect individual risks, can provide the right incentives for property owners, but it can be politically difficult to implement.

Pricing resources can be as complicated as pricing risk. In the electric power sector, for example, the fact that we pay time-averaged rates means that no one has an incentive to conserve during peak periods, which will become ever more problematic with the rising use of air-conditioning. On the other hand, average cost pricing allows utilities to smooth a lot of the price variability that might result from extreme weather (e.g., costs may spike in June, but averaging over a year could expunge most of this variation). There may be no clear case for either, but choosing one approach over the other will inevitably constrain the universe of adaptation options.

Clearly, it is difficult to estimate both the total cost of climate damage and how it is distributed across different stakeholders, but this information is essential for guiding both public and private decisions. Determining the other side of the net benefit calculation, the cost of adaptation, is equally challenging. High climate damage costs may arouse people and governments to action, but the appropriate adaptation response will have to be informed by an understanding of what options and resources are available to mitigate adverse climate impacts. Although this study focuses on climate damage estimates, we summarize the main perspectives on measuring adaptation cost for the reader's reference.

Many governments and multilateral institutions have invested substantial resources to better understand adaptation. Four leading examples of the latter are the Intergovernmental Panel on Climate Change (IPCC), the United Nations Environmental Programme (UNEP), the Organisation for Economic Co-operation and Development (OECD), and the World Bank. To illustrate the complexity of adaptation assessment, below is a summary of how each of these institutions defines adaptation cost:

IPCC—Adaptation costs are the costs of planning, preparing for, facilitating, and implementing adaptation measures, including transaction costs.[1]

UNEP—The cost of adaptation is the investment required in adaptation measures aimed at minimizing the damages from future climate hazards.[2]

OECD—The cost of adapting to climate change is the sum of investment costs and operating costs linked to the establishment of adaptation strategies.[3]

World Bank—Corresponding to a chosen level of adaptation is an operational definition of adaptation costs. If the policy objective is to adapt fully, the cost of adaptation can be defined as the minimum cost of adaptation initiatives to restore welfare to levels prevailing before climate change. Restoring welfare may be prohibitively costly, however, and policy makers may opt for an efficient level of adaptation instead. Adaptation costs would then be defined as the cost of actions that satisfy the criterion that their marginal benefits exceed their marginal costs. Because welfare would not be fully restored, there would be residual damage from climate change after allowing for adaptation.[4]

Comparing these definitions, even superficially, suggests that they could yield quite different rankings among adaptation alternatives. This reminds us that investments in soft as well as hard infrastructure will be needed for effective response to climate change. Clearly, institutions themselves need to adapt and promote more coherent decision making and dialogue in the face of a momentous emergent challenge to the public interest.

Two final points are worth emphasizing in the context of climate change costs. First, the disaster assessments associated with extreme weather and other events remind us that, in the context of natural processes, the loss function is quite asymmetric. For example, a hurricane predicted to be a major storm may pass by with very limited long-term damage, or it may completely devastate communities. This asymmetry results from threshold processes (for example, flooding, structural failure, and evacuation orders), and it renders risk-adjusted average costs of limited use because they overestimate the cost of most events and underestimate the cost of a catastrophic minority.

Finally, despite much progress that has been made in environmental economics, the value of environmental services from baseline resources still plays only a minor role in climate assessment. We are aware, of course, of our dependence on many amenities of today's world, but we have quite limited means of including these valuations in the baseline comparison for the cost of inaction, let alone the value of set priorities for restoring environmental services as part of the adaptation process. Much more progress is needed if we are to reliably factor these services into a more comprehensive cost-benefit analysis of climate policy.

TIMING

In its most general form, climate adaptation can be seen as a form of insurance, incurring (adaptation) costs at one time to avoid (climate damage) costs at another time. Thus, as it does for other risk-based investments, timing plays a central role in adaptation decisions. Of course, with climate change, both the adaptation and damage costs are uncertain, and so is the primary behavioral variable, the discount rate that mediates these two costs in determining the net present value of an adaptation choice. Further complicating this situation is the fact that adaptation can be proactive or reactive, occurring before or after the onset of damages.

This multidimensional uncertainty is a serious impediment to decision making, individual or collective, but human institutions have remarkable capacity to manage complex risks, and financial markets have already actively engaged the climate issue through insurance, venture capital, and other channels. All these developments can be seen as tentative and even speculative, but available evidence, including evidence provided in this book, continues to suggest that very large financial stakes will depend on the course of both climate change and adaptation. Better data on expected costs is certain to strengthen the capacity of markets for hedging climate risk, just as it will support individual adaptation decisions and changes in behavior.

Uncertainty about timing renders adaptation decisions difficult and potentially more expensive to make, thus reinforcing the tendency to defer them. For example, Neumann et al. estimate an uncertainty premium on California seawalls. If we borrow money at 3 percent to build walls ten years before inundation, the project will cost 35 percent more than it would if we could build seawalls the moment they were needed. Extending the margin of safety to twenty years increases the premium to 81 percent of the nominal project cost. At an interest rate of 5 percent, the same safety margins would add 63 percent and 165 percent to project costs.[5] Because sea level is in fact highly variable, subject to the dynamics of waves, tides, and storms, we must accept some degree of uncertainty and its attendant costs, but the question remains: How much?

The third source of uncertainty, the discount rate, has been seen by many observers as a major obstacle to the progress of both the mitigation and adaptation agendas.[6] All agents have their own discount rates, and there are significant disparities between these rates.[7] Financial markets offer some help here, but averaging investor discount rates provides only limited guidance about the opportunity cost of capital and inter-temporal asset comparison. Market interest rates are adequate proxies for discount rates in many other public and private investment decisions, and we do not believe climate adaptation is any exception.

What is likely the greater barrier to financial commitment in this context is actual perception of risk-adjusted, discounted cost. Seismic risk offers a convenient metaphor. In the San Francisco Bay Area, for example, a new bridge is currently being built, even though its predecessor is still standing. This proactive commitment was made because seismic risk is real, and the decision is generally accepted because the public has internalized that risk. Why so with seismic risk but not climate risk? The reason relates to both credibility and timing. Every year residents around the state get gentle reminders that seismic risk is real and present, and thus we have a multibillion-dollar industry in earthquake engineering and retrofitting, but negligible activity in adaptation.

Climate risk may be just as real, but the public has not fully internalized this, and certainly it is not seen as a present threat. Most climate process research addresses the middle or end of the present century; indeed, much of the published research does not address events prior to 2030 and suggests that most of the damage from climate change would occur near the end of this century. This situation, however, is changing, as more recent evidence suggests that climate change may be occurring faster than originally predicted. More important, we are awakening to the fact that not only averages of natural variables (temperature, sea level, etc.) but variances are increasing as well. Sea level does not rise like water in a bathtub but varies from week to week, hour to hour, and minute to minute because of the combined action of water temperature, tides, and wave action. In turn, these are affected by random storms and other natural processes, with the result that sea level risk will present itself decades before the average sea level rises by one or two meters. Unlike changes in averages, changes in variance are very difficult to anticipate, and it may not be possible to identify and attribute these changes until long after they have begun.

FAIRNESS

Economic fairness can be seen from many perspectives, but the most direct one is probably the distribution of wealth. From this angle, climate change has significant implications for both wealth and fairness across California society. It will be apparent from the research in this book that climate threatens a diverse spectrum of economic assets, but it does so in ways that are unequal across stakeholders. Different parts of the state will experience temperature and sea level effects differently, for example. As with fire, this can make statewide risk pooling inefficient and inequitable, and it may even increase damages by promoting risk taking. Those with the means who invest in adaptation will suffer lower absolute, and perhaps net, costs, reinforcing inequality. Public investments in adaptation will likewise be differentiated in their incidence, benefiting some more than others.

How should these expenses be financed? When thinking about climate impacts, we might suppose that damage costs will fall on all three major players in the economy: households, enterprises, and government. In reality, however, the final bill will be passed through the markets and fiscal systems to households, which will face higher direct costs, combined with higher prices and taxes to cover costs incurred by the other two. The same logic, of course, applies to adaptation costs. The ultimate net costs may fall to the California household, but to estimate them is a very complex cost allocation problem. Our estimates only make piecemeal contributions, but a few general conceptual rules are worth keeping in mind. In a more socially efficient and fair world, costs of climate risk should be borne by those whose behavior gives rise to those risks. Likewise, adaptation costs should be born by those who benefit from the adaptations, whether they are publicly or privately financed.

Other metrics of fairness, such as economic vulnerability, occupational opportunity, and mobility, can all be examined from a climate change perspective. Some of these aspects are discussed in the sector chapters, but conceptual treatment of them is beyond the scope of the present discussion.

Lastly, it is important to recognize feedback between fairness issues and policy making. Distributional effects are very important determinants of policy itself. Whom public spending, goods, and services affect, how, and when are all intimately related to policy formation and implementation. Cost/risk averaging approaches that ignore this heterogeneity are unlikely to provide reliable guidance for the course of adaptation.

INCENTIVES

Because adaptation is a form of behavior, its economic characteristics can be better understood if we examine incentives. For example, if the present value of climate damages exceeds the costs of adaptation that can avert these costs, there is a clear incentive to adapt. In this simple framework, we need only wait until climate change costs reach critical levels and we can expect to see waves of spontaneous private agency working to limit these adverse effects. To the extent that we do not see this, we could assume adaptation costs are too high, individuals rationally discount expected damages for credibility and time, institutional or other barriers stand in the way, or individuals have countervailing incentives of some kind.

The last case is of particular interest, since unfortunately there are many reasons why individual expected costs are inconsistent with social costs. In the private sector, this kind of divergence usually arises from simple stakeholder diversity. For example, a coal company might well calculate a different net benefit from mitigation policy than a kindergarten. These disparities in private interest are commonplace and form a solid basis for public interest initiatives by government, which must implement policies that can redistribute net benefits to compensate adversely affected stakeholders and promote the greater good.

More difficult cases arise from policies that reinforce conflicting incentives or distort behavior. The most important form of such behavioral distortion is moral hazard, which in this context creates incentives to act in ways that contradict the intention of a policy. For example, fire insurance, if it is priced below the marginal cost of policyholders (i.e., the value of their individual fire risk), will promote risk taking. It has been known for at least a century that average cost fire risk pooling promotes this kind of adverse behavior, but this has not stopped the practice.[8]

The same logic applies to a larger universe of climate risks. Some examples, such as floodplain insurance, are obvious, while others, like public investments to protect beaches, are more subtle. In broader financial markets, moral hazard can be very difficult to eliminate because of public fiduciary responsibility. According to the World Bank, of the nearly one hundred banking crises that have occurred internationally during the last twenty years, all were resolved by bailouts at taxpayer expense. Obviously, this precedence seriously compounds moral hazard. In the case of catastrophic risk insurance, such as that which may accompany escalating climate risk, the same risk of policy distortion is likely to emerge.

Incentives are usually necessary but not sufficient to explain behavior. In the same example, this means that agents will adapt if they can, if they have the resources to invest or can get them through accommodating capital markets. Of course they also need the information to understand both their incentives and adaptation options, and institutions must not obstruct their path to adaptation. Thus we see that sufficient conditions for adaptation may be much more complex than the simple logic of behavior would suggest, and actual capacity to adapt will depend on institutions as well as individuals.

INSTITUTIONS

If the world were organized by Adam Smith, individual agents would make individual decisions based on individual information, acting on this information autonomously and striving for self-interested profit yet achieving efficient overall resource allocation and many benefits for society as a whole. In these ideal circumstances, there would be no need for collective action or agencies such as governments, producer groups, or labor unions to achieve efficiency. Obviously, we do not live in such a world, and institutions influence nearly every dimension of our lives. The significance of this for climate adaptation is that institutions may facilitate or retard our progress toward a more sustainable economy, reducing or even increasing the costs of climate change and/or adaptation.

Like the other conceptual topics discussed in this chapter, institutional economics occupies academic volumes of its own, so we can only touch on salient features related to climate risk. As has already been mentioned, adaptation is essentially a local issue, while climate change is global. This means that governmental institutions, which span these domains, need some degree of consistency in responding to climate change. Unfortunately, the institutional landscape relevant to climate, mainly resource and regulatory agencies, is very fragmented and has only the most limited policy coordination. This is actually true at most levels of government in the United States, and California is no exception. For example, clean air standards are state policies, while fuel economy is national. Forestry, fishery, and agricultural regulations are largely national, while land use, water supply, fire management, and public health policies are controlled at the county or city level.

Another major institutional challenge is the political economy of interest groups. In Adam Smith's work all economic agents have equal, almost inaudible voices. The real world, of course, is very different, and concentrations of economic interests almost always coincide with concentration of political influence. To the extent that such groups have short-term goals that diverge from effective long-term adaptation, the larger public interest may be undermined. For example, developers might push local governments for development rights in high flood risk areas, as neither is liable for flood damages. This problem need not be seen as an ethical failure but can simply be considered a social distortion that applies different weight to different stakeholder interests. Our political system is sometimes interpreted as assigning equal per capita weights to things such as welfare, opportunity, and entitlement, but the real economy often recasts this through political advocacy financed by stakeholder investment. In the context of climate damage, the most important implications of this problem are structural bias in favor of industry interests and short-term growth objectives. Both of these have historically contributed to higher expected climate risk and moral hazard with respect to the magnitude and ultimate incidence of climate damage costs.

A third important institutional issue relates to the behavior of officeholders, those who help formulate, implement, and enforce policies related to climate change and adaptation. There are fundamental challenges to aligning the behavior of officeholders with longer-term, more inclusive perspectives of present and future generations. Climate change may be happening faster than many people believe, and climate damage may come sooner than most expect, but there are very few predictions of significant adversity that fall within the next election cycle for any officeholder in California or Washington. For this reason, even though action today might reduce adaptation costs for the majority of us, the result is unlikely to affect the next election. Officials are well aware of this, and climate policy is therefore subordinated to decisions that have more immediate impacts, even though these impacts might ultimately be much smaller.

More generally, we can see that, because their tenure is limited, officeholders experience an inter-temporal version of moral hazard, focusing attention and resources on short-term public priorities at the expense of longer-term ones. Any uncertainty about the long-term consequences of their actions reinforces this by weakening their accountability. This will be a problem as long as the evidence is not strong enough to link today's climate and adaptation decisions to tomorrow's consequences.[9]

CHAPTER TWO

Agriculture, Forestry, and Fishing

Agriculture, forestry, and fishing have long been fixtures in California's culture, politics, and economy. In the modern era, however, the three have a more limited economic role, accounting for about 1.5 percent of California's gross state product (GSP) and just over 3 percent of the state's labor force.[1] Despite this, agriculture, forestry, and fishing have important cultural, environmental, and economic values that are not captured by simple macroeconomic metrics. The nation's largest farm producer since the late 1940s,[2] California now accounts for nearly half of all U.S. fruit, nut, and vegetable production.[3] About one-third of California's land area is covered by forests,[4] and the contributions of these forests to ecosystem health and other qualities of life across the state are not easily monetized.

Climate change and climate policy will reshape California agriculture, forestry, and fisheries over the coming decades, and it is already having a negative impact on agricultural yields in other countries.[5] All three industries have evolved around historical climate conditions and are highly vulnerable to changes in the natural environment. Conversely, these sectors also have important influences on the environment. In addition to a wide spectrum of nonpecuniary environmental services, agriculture and forestry play important roles in the mitigation agenda. Agriculture is a significant source of greenhouse gas emissions,[6] but also, through biofuels, a potential low-carbon energy producer. Both the agriculture and forestry sectors are important sinks for carbon dioxide. Energy costs are a major share of expenditures for all three industries, and their enterprises will need to adapt to changing future energy prices and climate policies.[7] Because of their intimate relationship with the state's environment and natural resources, public interest and institutions will play an important role in facilitating their adaptation.

ECONOMIC IMPACTS OF CLIMATE CHANGE

At the lower range of predicted temperature increases, climate change could have positive or negative impacts on agriculture, forestry, and fisheries (table 1). Economically, these impacts occur first through changes in agricultural yield—mass per animal for livestock and output per unit area for crops, forest products, and fisheries. Changes in yields lead to price changes, which in turn affect producer decisions about what and how much to produce, as well as global supply and demand for agrofood and forest products. The adjustment process in agriculture, forestry, and fisheries, as prices change and producers and consumers all over the world respond, is extraordinarily complex, indeed more so than in any other sector examined in this study.

A growing body of research has examined the potential economic impacts of climate change on California agriculture and

TABLE I

Potential Positive and Negative Effects of Climate Change on Agriculture, Forestry, and Fishing

Climate change impact	Positive effect	Negative effect
Increase in temperatures	Earlier and longer growing season	Reduction in chill hours
	Earlier spring flowering	Loss of acreage for certain crops
	Increase in range for certain crops	Accelerated physiological development of crops
	Reduced frost	Desynchronization of flowering cycle and pollinator lifecycles
	Increased rate of photosynthesis	Increased irrigation needs resulting from higher evapotranspiration rates and reduction in soil moisture
	Increased growth rates for squid	Increase in wind erosion resulting from drier soil conditions
		Increased range and growth rates for pests, weeds, and pathogens
		Increase in plant CO_2 respiration
		Altered habitats for marine and anadromous fisheries
Higher CO_2 concentrations	Increased biomass production	
Increase in intensity and frequency of extreme weather		Early plant flowering
		Reduced pollination effectiveness
		Decreased ability for photosynthesis
		Increase in peak irrigation needs
		Increase in energy required to cool livestock
		Increases in crop damage from flooding and drought

Sea level rise	Increased flood risk
	Increased soil salinity
Increased seasonal water scarcity	Higher average water costs
Higher ozone levels	Decreased crop and timber yields
Changes in ocean pH	Changes in marine food webs and fish yields
Changes in ocean salinity and convection; changes in winds	Changes in physiology, development rates, reproduction, and survival of aquatic species
Increase in wildfires	Decreased timber yields

NOTE: For a review of impacts, see Dennis Baldocchi and Simon Wong, "An Assessment of the Impacts of Future CO_2 and Climate on Californian Agriculture," California Climate Change Center White Paper, CEC-500-2005-187-SF, 2006; Timothy Cavagnaro, Louise Jackson, and Kate Scow, "Climate Change: Challenges and Solutions for California Agricultural Landscapes," California Climate Change Center White Paper, CEC-500-2005-189-SF, 2006; Andrew Paul Gutierrez et al., "Analysis of Climate Effects on Agricultural Systems," California Climate Change Center White Paper, CEC-500-2005-188-SF, 2006; Gretta T. Pecl and George D. Jackson, "The Potential Impacts of Climate Change on Inshore Squid: Biology, Ecology and Fisheries," *Reviews in Fish Biology and Fisheries* 18 (2007): 373–85.

forestry, focusing on changes in average temperatures, pest populations and behavior, and, to a lesser extent, extreme weather (see tables 2–4). Much less attention has been given to the state's fisheries sector, where studies tend to be regional or global.

Owing to different modeling approaches, climate scenarios, time frames, and their coverage of climate impacts, there is disagreement among studies over whether climate change would have a net positive or negative aggregate economic impact on California farmers.[8] Comparing model results is indeed difficult. For instance, the models used by Deschênes and Greenstone and by Costello et al. hold prices and water availability constant. Howitt et al., allowing prices and water availability to change, argue that these two effects will be the main driving forces behind the agricultural sector's economic adjustment to climate change. For livestock, the impact of projected temperature increases is expected to be more unambiguously negative, with the burden of hotter summers generally outweighing the benefits of warmer winters.[9]

Current models in this field do not account for changes in the frequency, duration, and intensity of extreme weather events impacting agriculture.[10] Since 1995, extreme weather has had a modest effect on California's agricultural sector, leading to estimated damages of around $320 million per year (about 1 percent of annual agricultural revenues) and, in the two worst years, $1–2 billion in damages (4–5 percent of annual revenues) (fig. 2).[11] Damages from extreme weather most frequently have been from high rainfall events, though the two most costly extreme events over the past fifteen years were a freeze in 1998, mainly affecting fruit growers, and a heat wave in 2006, mainly affecting the dairy industry.[12] Rising temperatures would reduce freezing risk but

TABLE 2

Projected Economic Impacts of Climate Change on California Crops

Region	Climate model–emissions scenario	Time frame	Findings	Study
Statewide	HadCM2-IS92A	2070–2099	15% decrease in agricultural profits	Deschênes and Greenstone
Sacramento, San Joaquin, Tulare, Southern California	Warm-dry (climate model and emissions scenario not specified)	2050	11% decrease in farm revenues 20% reduction in cropped acreage	Howitt et al.
Statewide	PCM-B1	2010–2039	5%–11% increase in agricultural profits	Costello et al.
		2040–2069	6%–12% increase in agricultural profits	
		2070–2099	8%–16% increase in agricultural profits	
	PCM-A2	2010–2039	4%–8% increase in agricultural profits	
		2040–2069	12%–25% increase in agricultural profits	
		2070–2099	16%–36% increase in agricultural profits	

NOTE: HadCM2 is a medium-sensitivity and PCM a low-sensitivity climate model. Climate sensitivity is the responsiveness of the earth's near-surface air temperature to changes in radiative forcing. B1 is a low greenhouse gas emissions scenario; IS92A and A2 are business-as-usual emissions scenarios.

SOURCES: Olivier Deschênes and Michael Greenstone, "The Economic Impacts of Climate Change: Evidence from Agricultural Output and Random Fluctuations in Weather," *American Economic Review* 97 (2007): 354–85; Richard Howitt, Josué Medellín-Azuara, and Duncan MacEwan, "Estimating the Economic Impacts of Agricultural Yield Related Changes for California," California Climate Change Center, CEC-500-2009-042-F, 2009; Christopher J. Costello, Olivier Deschênes, and Charles D. Kolstad, "Economic Impacts of Climate Change on California Agriculture," California Climate Change Center, CEC-500-2009-043-F, 2009.

TABLE 3

Projected Economic Impacts of Climate Change on California Dairy

Climate model–emissions scenario	Time frame	Reduction in output
PCM-B1	2020–2049	0%–2%
	2070–2099	0%–7%
HadCM3-B1	2020–2049	0%–2%
	2070–2099	3%–10%
PCM-A1fi	2020–2049	0%–2%
	2070–2099	2%–11%
HadCM3-A1fi	2020–2049	0%–4%
	2070–2099	8%–22%

NOTE: HadCM3 is a medium-sensitivity and PCM a low-sensitivity climate model. B1 is a low emissions scenario; A1fi is a high emissions scenario.

SOURCE: Katharine Hayhoe et al., "Emissions Pathways, Climate Change, and Impacts on California," Proceedings of the National Academy of Sciences 101 (2004): 12426.

increase the frequency and intensity of heat waves, with uncertain impacts on rainfall.

A number of other impacts lie outside the scope of current models, including the effects of pests and ozone. Rising temperatures are expected to have a range of effects on insects, weeds, and pathogens, including increased range and population growth rates.[13] In tandem with a drought that weakens plant defenses, major pest outbreaks could devastate the agricultural sector, but these complex interactions are difficult to capture with models. Changes in temperatures and atmospheric chemistry will frustrate ongoing efforts to reduce ozone levels (see chapter 8), but the impacts on agriculture depend upon the

TABLE 4

Projected Economic Impacts of Climate Change on California Forestry

Climate model–emissions scenario	Time frame	Findings	Study
HadCM2 and PCM, range of temperature scenarios	2100	$0.1–$1 billion (NPV) loss for producers $13–$14 billion (NPV) gain for consumers	Mendelsohn
PCM-A2	2020	(−3.2%)–(−3.5%) change in timber value	Hannah et al.
	2050	(−3.3%)–(−3.6%) change in timber value	
	2080	(−8.1%)–(−8.5%) change in timber value	
GFDL–A2	2020	(+0.2%)–(+0.5%) change in timber value	
	2050	(+0%)–(+0.4%) change in timber value	
	2080	(−4.7%)–(−4.9%) change in timber value	

NOTE: HadCM2 is a medium-sensitivity climate model; PCM and GFDL are low-sensitivity climate models. B1 is a low emissions scenario; A2 is a business-as-usual emissions scenario. NPV is net present value.

SOURCES: Robert Mendelsohn, "A California Model of Climate Change Impacts on Timber Markets," Appendix XII in T. Wilson et al., *Global Climate Change and California: Potential Implications for Ecosystems, Health, and the Economy,* PIER Publication No. 500-03-058CF, 2003; Lee Hannah et al., "The Impact of Climate Change on California Timberlands," California Climate Change Center Final Paper, CEC-500-2009-045-F, 2009.

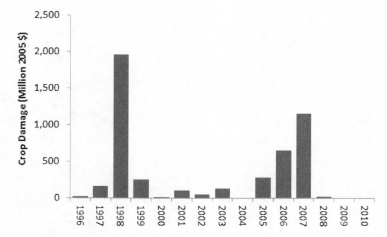

Figure 2. Crop damages from extreme weather in California, 1996–2010. Source: Data are from NOAA, Office of Climate, Water, and Weather Services, "Natural Hazard Statistics," www.nws.noaa.gov/om/hazstats.shtml.

complex interaction between mitigation policies, other changes in climate, and ozone levels. As a reference, the current cost of ozone impacts on California agriculture is likely on the order of several hundred million dollars.[14]

Although there is disagreement over net economic impacts, all studies agree that the effects of climate change will be uneven across crops.[15] For instance, increases in average temperatures would attenuate the negative impacts of frost, benefiting some crops, but it would also reduce the number of chill hours needed by certain fruit and nut trees to initiate flowering. For a diversified agricultural sector, aggregate adjustment to climate change might be relatively smooth, but California agriculture, despite producing a wide array of goods, is highly concentrated in terms of both value and location. Just ten products accounted for more than 70 percent of California agricultural output in 2009 (fig. 3),

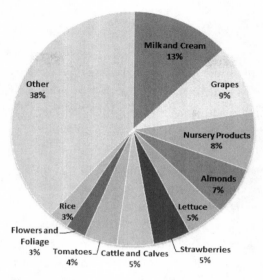

Figure 3. Shares of California agriculture output value in 2009. Source: Data are from California Department of Food and Agriculture, "Agricultural Statistical Review," California Agricultural Resource Directory 2010–2011, 2011.

and fifty-six percent of the value of the state's agricultural output was produced in five of its fifty-eight counties in 2009.[16]

As in agriculture, the net economic impacts of climate change on California's forestry sector remain uncertain. Two studies by Battle et al. on how climate change affects pine forests, for instance, produced contradictory results, with one study predicting a decline and the second predicting an increase in tree volumes.[17] Worst-case scenarios, in which drought is combined with the spread of pathogens, could lead to tree mortality on a large scale, as may have occurred in the late Permian period.[18] Mendelsohn argues that the largest climate impact on California's forestry sector would be price-driven, with a net loss to producers but a net gain to consumers as timber prices fall with increases in yields and harvests.[19] Hannah et al. predict that impacts will be highly uneven across counties, with timber value increasing in some counties while falling in others.[20] Like the agricultural sector, California's forestry sector is highly concentrated geographically, with the top five counties producing 58 percent of the state's timber in 2007.[21]

The leading fishery species, squid and salmon, will both be affected by climate change, but potential impacts are still poorly understood. Higher seawater temperatures could increase squid growth rates, but they could also lead to changes in size, population structure, and metabolism that would make squid populations more vulnerable.[22] Changes in squid physiology could lead to changes in the location and timing of spawning, with implications for both the fishing industry and natural resource managers.[23] More frequent temperature extremes could also impact availability of squid, salmon, and pelagic fish populations. In the 1997–98 El Niño–Southern Oscillation (ENSO) season, squid landings plummeted from their then record high of 110,000 metric tons in 1996–97 to less than 1,000 metric tons because of the high temperatures associated with El Niño.[24]

Climate change will also affect salmon by altering their anadromous (freshwater/saltwater) habitats and migration between them. Earlier snowmelts could push salmon into the ocean earlier, without sufficient spring phytoplankton to feed them. The balance between predators and baitfish populations at that time could affect salmon survival rates by altering the number of predators.[25] The different phases of the Pacific Decadal Oscillation (PDO), an interdecadal climate phenomenon in the Pacific, also appear to have an influence on salmon stocks, with dwindling salmon populations during warm phases and large populations during cool phases.[26]

Changes in both climate averages and extremes will contribute to spatial and temporal shifts in water availability, which may be the most important climate impact for agriculture, forests, and freshwater fishing. However, given that agricultural and environmental water use are intimately linked to urban water use, we discuss potential impacts of changes in water supplies in chapter 3 and focus here on nonwater aspects of climate change impacts.

ADAPTATION OPTIONS AND CHALLENGES

In principle, agriculture, forestry, and farm fisheries in California could be highly adaptive and resilient. With more reliable long-term climate forecasts, crop switching, breeding, and improved management practices could sustain profits as agroecological regions shift. New varieties of fruits and nuts that require less chill time could increase survival rates as chill hours decline. Better monitoring and understanding of pests, weeds, and diseases could lead to improved control and reduced damages. More efficient water use and allocation (e.g., trading) could reduce farmers' economic exposure to drought.

Improving adaptive capacity in these sectors will require striking a balance between private sector adaptation and public sector research and extension assistance. Facilitating a significant degree of private adaptation, by bringing input and output prices more in line with market prices, will allow production to adjust to long-term changes in growing conditions. However, many adaptation measures—information and monitoring in particular—have public goods characteristics and will require sustained government support. Providing more public resources to agriculture and forestry would require reversing a decades-long trend of scaling back public assistance to these sectors, such as the Cooperative Extension System. Enhanced public support may be necessary to deliver adaptation capacity to the next generation of farmers.[27]

Dealing with the impacts of longer-term, gradual changes in climate may be simpler than dealing with changes in climate variability. Insurance is the primary tool to protect producers from weather disasters. In the agricultural sector, insurance is provided at subsidized rates through the Federal Crop Insurance Corporation. Because of the high risk involved, private markets for crop insurance do not exist in the United States. The forestry sector, on the other hand, has a more dynamic insurance and reinsurance market. The future of these insurance markets in an era of climate risk is uncertain, but it will have important implications for sustaining investment in agriculture and food security. Although public insurance is a behavioral minefield (see chapter 7), public institutions will probably have to make sustained commitments here. Because of California's relatively high vulnerability and the fact that it does not produce staple commodity crops, the state government may have to supplement federal program support.

Adaptation at a state level in California will have to account for both spatial heterogeneity in climate impacts and the uneven importance of the agricultural and forestry sectors across different counties (figs. 4 and 5). For instance, although agriculture accounts for less than 3 percent of the state's labor force, in nine of California's fifty-eight counties it accounts for more than 15 percent of civilian employment.[28] Although the forestry sector is a very small part of California's gross state product, the for-

Figure 4. Share of agricultural employment in
total employment. Source: Data are from California
Department of Finance, "California County Profiles," www.dof.ca.gov/HTML/FS_DATA/profiles/
pf_home.php.

est products industry is concentrated in the northern part of the
state. Policies and funding for agricultural and forestry adaptation at a state level should be designed to account for differences
in climate risk and economic vulnerability at a county level.

Climate change damages to agriculture, forestry, and fisheries sectors could extend beyond direct monetary impacts on the
private sector and local governments. For instance, forestland
in California is about evenly split between public and private
ownership.[29] Revenues from timber harvested on public lands
are used to fund forest management. Without these revenues,

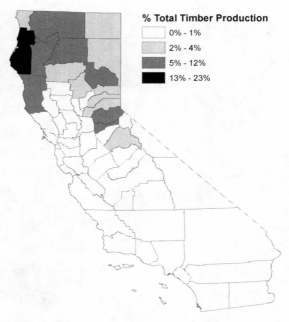

Figure 5. Shares of total California timber production. Source: Data are from California Department of Finance, "California County Profiles," www.dof.ca .gov/HTML/FS_DATA/profiles/pf_home.php.

forest agencies might be forced to scale back activities, which would have adverse implications for wildfire and pest management. Understanding where these connections lie and securing longer-term funding for public services will be important dimensions of climate adaptation across all economic activities.

In summary, despite their small contribution to aggregate state economic activity and employment, agriculture, forestry, and fishing are very important politically and are deeply linked to a resource base (water, land) that is threatened by climate change.

Climate change portends significant impacts on all thee activities in California. In lower warming scenarios, some of these changes will be beneficial for agriculture and forestry, although there is still a debate about the net impact. Agriculture will experience at least seasonal and perhaps annual water scarcity, spatial changes in crop eligibility, higher yields and pest activity, and greater vulnerability to energy prices (agrofuel and chemicals). Forestry will experience higher yields but also elevated fire risk and drought vulnerability. Fishery changes are less predictable, but rising sea temperatures may displace coastal fisheries. Yield benefits, however, are unlikely to offset high adjustment and transaction costs. Direct net cost estimates range from hundreds of millions to billions annually, although very little has been done to estimate the very extensive costs of sector displacement, food source substitution, and land value effects. Structural change in these sectors is expensive because of high fixed-asset and fixed-cost proportions. For this reason, higher-quality information that supports early action can play an essential role in limiting adjustment costs. There are many activity-specific risk management and adaptation options open to agricultural operators, but they need reliable guidance to commit to them. Forestry is a regional activity in the state, and, over the time horizon considered here, demographic trends will probably exert more influence on the forestry sector than climate change. For fisheries, better-quality information and risk management tools (e.g. insurance) are again essential to facilitate adjustment.

CHAPTER THREE

Water

California's history over the last two centuries has been inextricably intertwined with water scarcity and rivalry, challenges made inevitable by the uneven temporal and spatial distribution of the state's water supply and demand. Most of California's water arrives as winter rains that feed rivers in the north, whereas water demand is highest during late spring and summer in the southern part of the state. Compounding this problem is high year-to-year variability in precipitation, with California regularly oscillating between droughts and floods. To overcome these spatial and temporal mismatches, the state has developed an extensive system of water storage and conveyance, much of which was designed and built more than half a century ago.

California's water supply is highly vulnerable to changes in climate. In the eastern part of the state, snowpack in the Sierra Nevada Mountains is the state's second largest water storage facility after groundwater, releasing water during California's dry summers. With a rise in average temperatures, some or even most of this snowpack could disappear, intensifying spring floods and threatening summer drought.[1] From the west, increased storm severity and rising sea levels threaten the Sacramento–San Joaquin Delta, an inland river delta east of the San Francisco Bay that supplies 15 percent of the state's total water needs.[2] Climate change impacts on these and other parts of California's natural and engineered water infrastructure pose important risks for water supplies.

Reductions and increased variation in water availability would strain the shifting and already contentious balance among California's agricultural, environmental, and urban water users. The environment—water allocated to protect ecosystems—currently accounts for nearly half of California's total water use, but further restoring degraded ecosystems would require more water for environmental use.[3] Urban users account for only around 10 percent of total water use, but continued population growth means that this share will likely grow. Agriculture, the remaining 40 percent of water use, has seen its share steadily decline since the 1980s and faces an uncertain water future.[4] The full costs of adapting California's water sector to a changing climate, as we describe below, are very much bound up with this political economy of water and the institutions required to effectively manage it. Engineering solutions to water scarcity may ultimately seem modest compared to the political and social resources that will needed to change patterns of water allocation and use.

ECONOMIC IMPACTS OF CLIMATE CHANGE

Seasonal patterns of water availability in California, and the infrastructure to store and move water, necessarily evolved around climate conditions. Initial conditions in the state required extensive water infrastructure to accommodate demographic trends and agricultural development, but climate change will impose new adaptation stresses. Rising temperatures are expected to reduce the storage capacity of the Sierra snowpack by shifting winter precipitation from snow to rain and advancing the onset of spring snowmelt. This shift would lead to higher winter river flows and lower flows during the late spring and summer, increasing the likelihood and severity of winter floods and summer droughts.

Sea level rise is expected to lead to saline intrusion in the Sacramento–San Joaquin Delta, as well as other coastal rivers and aquifers. By changing tides and storm surges, sea level rise could also threaten the extensive levee system that protects the Sacramento–San Joaquin Delta and other coastal areas from flooding. Although these impacts, and a number of others (see table 5), are more certain, timing such damages is very difficult because of the agency of storm, tidal, and wave activity. In this sense, coastal damage risk more closely resembles seismic risk, and defensive measures should likewise be timed with considerable foresight.

From the water availability perspective, there is no consensus yet on how a warming climate would affect future rainfall patterns in California, and this significantly complicates responsible planning for water storage and conveyance capacity.[5]

Because of its role in smoothing seasonal water availability in California, the Sierra snowpack has been particularly important to California's water sector. Even in scenarios where dangerous climate change is avoided, however, the snowpack could shrink by 30 to 70 percent by the end of the century (fig. 6). In higher warming climate scenarios, 60 to 90 percent of the snow-

TABLE 5

Potential Climate Change Impacts on California's Water Supplies

Direct physical change	Impact
Increase in winter rainfall	Increase in flood probability Change in water supply
Earlier melting of snowpack	Increase in flood probability Decrease in summer water supply
Decrease in snowpack extent	Decrease in summer water supply Increase in drought probability
Greater variability in precipitation	Increase in frequency, duration, and intensity of droughts and floods
Sea level rise	Increase in probability of levee failure and flooding Saline intrusion in the Sacramento–San Joaquin Delta and coastal aquifers
More intense coastal storms	Inundation of water treatment facilities
Higher temperatures	Increase in water temperatures, changes in habitat
Increase in evapotranspiration	Increase in water demand for agriculture and landscaping

pack could disappear by century's end. Climate models are somewhat more consistent in their longer-term projections of a warming climate on the Sierra snowpack, but the near term is highly uncertain. In wetter climate models the snowpack actually increases over the next two decades, whereas in drier climate models it could shrink by a third.

Because of its economic and environmental importance, California's water sector has been the most extensively studied of all the sectors that may be seriously impacted by climate change.

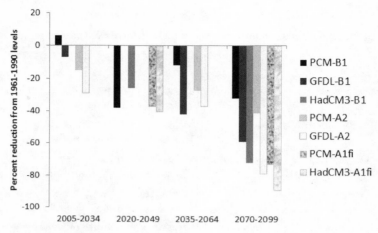

Figure 6. Changes in the Sierra snowpack under different climate models and emissions scenarios. PCM, GFDL, and HadCM3 are climate models, and B1, A2, and A1fi are emissions scenarios. Sources: PCM-B1 2020–2049, all PCM-A1fi, and all HadCM3 results are from Katharine Hayhoe et al., "Emissions Pathways, Climate Change, and Impacts on California," *Proceedings of the National Academy of Sciences* 101 (2004): 12423. All other results are from Daniel R. Cayan et al., "Climate Change Scenarios for the California Region," *Climatic Change* 87 (2008): S37.

Studies focus mainly on two water-related climate change impacts: reductions in water supply that result directly from changes in hydrology and those that result from levee failure.

Also important but less researched are the inundation risks of Pacific storms on coastal wastewater treatment facilities, water quality impacts of saline intrusion in the Sacramento–San Joaquin Delta and coastal aquifers, and the effects of hydrological changes on river ecosystems. The last effect raises particularly challenging public policy questions because climate change will influence natural water distribution and withdrawal capacity on river ecosystems, as well as the public value of these ecosystems. These changes are very difficult to predict, but they have profound implications for property rights and equitable resource use.

Direct Reductions in Water Supply

With limited near-term flexibility in storage and allocation systems, significant reductions in water supply induce scarcity. In the short run, agricultural users, urban users, and government agencies in charge of environmental use can buy more water resources from each other or ration themselves. In the longer term, government agencies and water providers can invest on both the supply and demand sides, including water storage and conveyance infrastructure, desalination plants, end use efficiency, and water recycling and reuse infrastructure. These investments expand supply, but they do so at a cost.

The economic impacts of climate change–induced changes on water supply are complex, acting at the intersection of climate, hydrology, ecosystems, and human societies. Models can help to better navigate this complexity, but they can neither reduce nor reproduce it. Table 6 presents several cost projections of a climate change–induced reduction in California's water supplies.

The models used to generate the results in table 6 differ in their geographic coverage, climate scenarios, and assumptions, and these differences explain the significant range of results.[6] On average, however, the cost projections are relatively modest. If paid for only by the residential sector, for instance, a $369 million increase in annual operating costs by 2050 would raise water expenditures by about $1.50 per household per month.[7] The main driver of these cost increases, as illustrated by the Tanaka et al.

TABLE 6

Projected Economic Impacts of Climate Change on California's Water Sector

Time frame	Climate model–emissions scenario	Region	Incremental scarcity costs	Incremental operating costs	Source
2070–2099	GFDL-A2	Central Valley (agriculture), Southern California (urban)	$2,300 million/year[a]	n/a	Hanemann et al.
2080–2099	Hadley-IS92a (wet)	Sacramento Valley and Bay Delta, San Joaquin and South Bay, Tulare Basin, Southern California	–$237 million/year (2006 dollars)[b]	–$21 million/year (2006 dollars)	Tanaka et al.
	PCM-IS92a (dry)		$147 million/year (2006 dollars)	$1,663 million/year (2006 dollars)	
2050	GFDL-A2 (dry)		$121 million/year (2008 dollars)	$369 million/year (2008 dollars)	Medellín-Azuara et al. 2008
2050	GFDL-A2 (historical runoff)		$7 million/year (2008 dollars)	$17 million/year (2008 dollars)	Medellín-Azuara et al. 2009
	GFDL-A2 (dry)		$639 million/year (2008 dollars)	$264 million/year (2008 dollars)	

NOTE: Hadley is a medium-sensitivity climate model; PCM and GFDL are low-sensitivity climate models. Climate sensitivity is the responsiveness of the earth's near-surface air temperature to changes in radiative forcing. IS92A and A2 are business-as-usual greenhouse gas emissions scenarios. Where reported by the authors, precipitation scenarios are shown in parentheses.

SOURCES: Michael Hanemann et al., "The Economic Cost of Climate Change Impact on California Water: A Scenario Analysis," PIER Project Report, CEC-500-2006-003, 2006; Stacy K. Tanaka et al., "Climate Warming and Water Management Adaptation for California," *Climatic Change* 76 (2006): 361–87; Josué Medellín-Azuara et al., "Adaptability and Adaptations of California's Water Supply System to Dry Climate Warming," *Climatic Change* 87 (2008): S75–S90; Josué Medellín-Azuara et al., "Water Management Adaptation with Climate Change," California Climate Change Center Final Paper, CEC-500-2009-049-F, 2009.

[a] The estimates in Hanemann et al. are for net revenue losses for agricultural users (2004 dollars) and replacement costs and loss of consumer surplus for urban users (2003 dollars and base year unspecified, respectively), which are conceptually different from the scarcity cost metrics used by Tanaka et al. and Medellín-Azuara et al. Because the studies seek to answer the same question, however, it is useful to compare the results.

[b] Negative scarcity costs imply that there is a surplus.

and Medellín-Azuara et al. (2009) results in table 6, is changes in precipitation patterns rather than rising temperatures.

The economic impacts of water scarcity depend crucially on whether water can be transferred from agricultural to urban users. If water transfers are more flexible, as assumed in the Tanaka et al. and Medellín-Azuara et al. studies, scarcity costs range from small to nonexistent. Without that flexibility, as assumed in the Hanemann et al. study, scarcity costs are much higher. In the Hanemann et al. study, the urban users in Southern California bear the lion's share of scarcity costs in the driest years, and large-scale rationing becomes necessary. With flexible water transfers, the agricultural sector bears most of water scarcity costs because urban users have a higher willingness to pay for water. Without flexibility, urban users shoulder the larger burden because they account for most of the growth in water demand.

Beyond its impact on water utility revenues, rationing water use in the residential sector reduces individual utility but has small direct macroeconomic impacts. Individual utility, additionally, may be unsatisfactory as a basis for water policy decisions, as the doctrine of "reasonable use" governing water allocation in California implies that willingness to pay should not, by itself, be a guide to water allocation.

None of the cost metrics used in the studies in table 6 has a direct relationship to traditional macroeconomic indicators, such as gross state product or employment. Over the last century, economic growth in California has essentially been delinked from water consumption. The state's economy grew by 3.6 percent per year from 1980 to 2005, while total gross water use fell by 0.6 percent per year.[8] The sustainability of both trends, however, and the effects on the California economy of larger and longer-term reductions in water supply or of more frequent and severe short-term water scarcity remains uncertain.

Outside agriculture, commercial business is a small water consumer in California, accounting for around 20 percent of urban water use and just 2 percent of total water use.[9] Small aggregate water use conceals the fact that some businesses, such as food processors, chip manufacturers, and energy producers, are water intensive, and many important downstream sectors depend on these industries.

The studies listed in table 6 also highlight challenges that California's water sector faces even without considering climate change. Population growth is foremost among these challenges. The California Department of Finance estimates that the state's population will reach 59.5 million people by 2050, an increase of 20.4 million people from 2010.[10] At current levels of per capita water use, population growth of this magnitude would require expanding total human water supplies by almost 15 percent over current levels.[11] Tanaka et al. argue that the water scarcity costs resulting from population growth could far exceed those from climate change over the long run.[12]

There is also evidence that adapting to natural climate variability will raise the cost of water supply in California. Paleoclimatic research has revealed evidence of droughts lasting for as a long as a century in medieval California (ca. 900 to 1350 CE).[13] Longer-term droughts, lasting from 20 to 100 years, appear to have occurred at intervals of 80 to 230 years in the late Holocene (ca. 1000 BCE to 1700 CE).[14] For reference, no California drought in the last century has lasted for more than six years without an intervening year of above-average precipitation.[15]

TABLE 7

Projected Economic Impacts of Levee Failure
on the Central Valley Water Sector, 2077–2099

| | Total scarcity costs | |
Precipitation	*No import constraints*	*Import constraints*
Before drought	$4 billion	$12 billion
After drought	$2 billion	$14 billion
Wet year	$0.2 billion	$10 billion

NOTE: These projections are based on hydrological conditions from the low-sensitivity GFDL climate model, using the business as usual A2 emissions scenario.

SOURCE: Sebastian Vicuña, Michael Hanemann, and Larry Dale, "Economic Impacts of Delta Levee Failure Due to Climate Change: A Scenario Analysis," PIER Project Report, CEC-500-2006-004, 2006.

Levee Failure

The Sacramento–San Joaquin Delta, protected by thousands of miles of levees, supplies water for agricultural and urban users in the Central Valley and Southern California through the State Water Project and Central Valley Project. Sea level rise, storms, and an increase in winter runoff have the potential to inundate the delta's aging levee infrastructure, disrupting water deliveries throughout California on a large scale. Levee failure could allow brackish water from the San Francisco Bay to enter the water supply, affecting water quality and temporarily reducing water supplies.

The economic impacts of flooding and saline intrusion on the water sector result from temporary reductions in water supply and depend on precipitation conditions when a flood event occurs.[16] If levee failure were to occur before or after a drought, water supplies would already be tight and water transfers would be less likely, leaving the system with little flexibility. As Vicuña et al. demonstrate (see table 7), in wet years levee failure may cause limited damage. More important, though, is the extent of an area's water supplies at risk. If for instance, in addition to its State Water Project allocation, urban Southern California needed to secure water from delta sources that are also at risk from levee failure, Vicuña et al. estimate that the costs of levee failure would rise by an order of magnitude.

ADAPTATION OPTIONS AND CHALLENGES

With a long history of adapting to changing conditions of water supply and demand, California's water sector is, in some ways, well positioned to adapt to climate change. Water management institutions have guided the state's water system through major droughts and have sustained population growth and a significant shift in management practices toward balancing human use with protecting river ecosystems.[17] The state's economy and agricultural sector have also demonstrated an ability to deal with variability in water supplies, and both have the capacity to adjust further. The economy can make progress through scaling up end-use efficiency and water transfers, and the agricultural sector can shift toward higher-value crops and water-efficient technologies.[18]

In other ways, California's water sector is headed for a momentous and potentially disastrous reckoning. Water management in California over the past four decades has been characterized by conflicts among stakeholders that have diminished their capacity to solve problems. For instance, although the Sacramento–San Joaquin Delta is on the verge of physical and biological collapse,[19] efforts to reconcile competing interests and solve the delta's complex natural and social problems have been futile.[20] This failure is an ominous precedent for California's public and private water institutions as they face the complex management and allocation challenges that climate change will present.

California's water infrastructure and institutions, designed for an era of greater perceived certainty, will also need to adapt to a changing future. Currently, water resource planning is based on historical hydrologic conditions spanning less than a century's worth of data—an assumption known as the "stationarity hypothesis."[21] As explained above, climate variability over the last century was probably low relative to the rest of recent human history, which suggests that California's water system is not prepared to handle longer-term natural climate variability, let alone the compounding uncertainties of a changing climate.

Even in a worst-case climate scenario, California's businesses, ecosystems, and residents will continue to have access to water. The question is, at what economic and environmental costs? Solutions for adapting the state's water infrastructure and institutions range from more capital-intensive, large-scale options such as desalination plants and new reservoir capacity to lower-cost but more institutionally complex options such as joint aquifer-reservoir management, water trading, reoptimizing reservoir operations, end-use efficiency, and water pricing. Without well-functioning institutions and constructive policy dialogue, low-cost adaptation solutions are less likely.

In the institutional sphere, two interrelated challenges are most prominent for California's water sector: how to facilitate collective action and how to modernize the underlying legal and regulatory frameworks. Organizationally, water management in

California is extraordinarily complex, with hundreds of public and private water utilities and irrigation districts, state government agencies, and federal government agencies deployed across a wide and overlapping array of regulatory and commercial roles. Although decentralization and organizational diversity have encouraged local innovation, they make it difficult to achieve the coordination and cooperation required for lower-cost regional planning that coherently integrates water supply, flood management, and ecosystem conservation.[22]

California has a byzantine patchwork of laws and regulations that govern water management, allocation, and use, many of which are not well suited to an era of greater climate variability. Surface and groundwater rights, for instance, are still separately administered in many areas, and who actually owns the rights to groundwater is often unclear.[23] Conjunctive management of surface and groundwater, where runoff is stored in aquifers in wet years for use in dry years, would require a clearer and more systematic network of rights to achieve anything approaching efficient temporal and spatial allocation. Modernizing laws and regulations will likely take decades, particularly because it is not even clear in some cases which laws and regulations need to be changed.[24]

All of the barriers to better water management mentioned above suggest the need for strong public sector leadership, coordination, and intervention, and illustrate the primacy of institutions in determining water management outcomes. These barriers also suggest that business-as-usual water allocation is unlikely to resemble what a single public sector water planner might consider optimal. Climate change will severely test this divergence between private and public interest, and the institutional costs of adaptation will potentially be very large.

Thus we see that water is essential to all economic activities, but it is very unequally distributed across California. In-state water supplies originate in northern weather patterns, yet most water is consumed in the south because it is relatively more arid and densely populated. Because of this, extensive water reallocation has been necessary to support the state's historic economic growth patterns. Meanwhile, the primary functional division in water use has been between agriculture and residential users, with industry using comparatively little water.

Most climate models agree that global warming will increase California's precipitation in winter and reduce it at other times. This may not change the state's annual water budget, but warming will dramatically reduce California's second most important water storage facility (after aquifers), the Sierra snowpack, leading to larger and more variable seasonal disparities in natural water availability. Combined with significant expected population growth, this will lead to considerable stress on existing water storage and allocation systems. Higher water flow variability will also lead to increased risks of flooding, saline intrusion, and drought-induced habitat destruction. Water conservation offers the most cost-effective means of reducing scarcity and its attendant costs, but it is unlikely to offset a substantial part of long-term growth in residential demand.

In the absence of climate defense measures, the potential costs of these climate impacts remain very uncertain, with estimates ranging from a few hundred million to several billion per year. To a significant extent, these differences are due to varying assumptions about how the state would adapt to scarcity. Initial conditions in the state's water economy are seriously distorted by legacy rights, allocation, and pricing policies, so there would seem to be scope for water trading systems to achieve more efficient allocation. In particular, urban water users pay about fifty times what agricultural (the heaviest) users pay, suggesting that markets could shift water in the event its scarcity value rises. Indeed, the leading simulation models used to study this assume that trading will significantly mitigate climate-induced scarcity. There are also out-of-state water resources, primarily from the Rocky Mountains, that are assumed to offer additional water at competitive prices. Finally, state groundwater resources are assumed to offer transitory flexibility to smooth annual water access.

We are concerned that institutional rigidities and Rocky Mountain water scarcity may lead water costs to escalate sharply in response to climate change. Trading systems are unlikely to operate smoothly with existing patterns of water entitlement and conveyance infrastructure.[25] Assuming the Rocky Mountains experience the same snow impacts as the Sierras, it is unrealistic to rely on this source as a backstop. Finally, California aquifers, the state's primary water storage facility, are not well captured by any existing models, and their entitlement and exploitation characteristics do not suggest that competitive allocation opportunities will emerge spontaneously in response to more acute scarcity.

Climate adaptation in the water context will require extensive investments in both hard and soft infrastructure. Public-private partnership can finance the former, but the latter will have to break new ground with respect to the public interest in water access and use. Defense or damage mitigation is feasible, beginning with existing renewal and replacement budgets. For longer-term adaptation, significant investments in storage, conveyance, and water management institutions are needed. We also believe

that more research will support a simple but challenging conclusion: California water has been too cheap for too long, and a
significant rise in its scarcity value could trigger intense rural-
urban competition and a complete reappraisal of rules governing
the state's water entitlements and private use.

CHAPTER FOUR

Energy

Energy is the prime mover behind California's diverse economy, powering everything from lab equipment in state-of-the-art research facilities, to transport and logistics that facilitate global trade, to kitchens in some of the world's finest restaurants.

Climate change has the potential to impact every component of California's energy supply chains, from resources and transport to conversion, distribution, and use. Government agencies, businesses, and households will also need to adapt the state's energy systems and use patterns to a changing climate at the same time that the state attempts to meet aggressive goals for reducing greenhouse gas emissions from energy use.

ECONOMIC IMPACTS
OF CLIMATE CHANGE

The list of potential climate change impacts on California's energy sector is long and varied, ranging from temperature effects on equipment to sea level rise and fuel transport (table

TABLE 8

Potential Climate Change Impacts on California's Energy Systems

Climate change impact	Potential energy system impact
Increase in average temperatures	Increase in air-conditioning and electricity use
	Reduction in winter heating
	Reduction in power capacity of transmission lines
	Reduction in efficiency of thermal and geothermal power plants
	Reduction in summer hydropower potential
	Changes in the amount, timing, and distribution of wind power
Increase in extreme temperatures	Increase in peak period air-conditioning and electricity use
	Increase in transformer failures
	Reduction in solar photovoltaic output
Increase in wildfires	More frequent (and perhaps prolonged) transmission outages
Increase in storm intensity and frequency	Increase in offshore rig risk, outages, and costs
	Increase in coastal power plant outages and costs
Changes in precipitation	Increase or decrease in winter hydropower availability

8).[1] Many of these impacts will occur according to timeframes that would allow incremental engineering solutions and infrastructure upgrades, replacement, or retirement. As long-lived energy infrastructure is upgraded or replaced, new designs can enable equipment to function across a larger range of operating conditions. In some cases, mitigation efforts could make current energy infrastructure obsolete by the time more serious climate impacts are expected to materialize. For instance, meeting California's 2050 greenhouse gas targets will likely require electrifying many uses of energy that currently rely on direct fossil fuel combustion, such as gasoline use in light-duty vehicles,[2] a transition that would decrease reliance on oil infrastructure.

The electricity sector is central to both mitigation and adaptation efforts. The two energy sector climate impacts that are more structural and less incremental, changes in hydropower availability and temperature-induced increases in electricity demand, are important electricity sector challenges.

Changes in the Amount and Timing of Hydropower Generation

Hydropower, brought from high-altitude rivers in the Sierra Nevada Mountains to coastal cities and the Central Valley over long-distance transmission lines, powered California's economic ascent during the late nineteenth century.[3] A century later, hydropower continues to be an important part of the state's electricity portfolio, accounting for between 9 to 30 percent of its total electricity consumption, depending on hydrologic conditions.[4]

A significant portion of California's hydropower consumption is imported, primarily from Columbia River dams in the Pacific Northwest and the Hoover Dam in Nevada. From 2002 to 2009, hydropower imports contributed an average of 6 percent of California's total gross system power and 29 percent of its hydropower consumption, with large interannual variability.[5] The timing of hydropower imports makes them a particularly important resource. Historically, as much as 10 percent of California's peak summer demand has been met with hydropower imports from the Pacific Northwest.[6]

Hydropower facilities are typically classified by head, or the height of the dam's water source, and water storage capacity. Low-head, high-storage facilities tend to be lower-elevation dams that are used for several complementary and/or competing purposes, such as flood control, recreation, irrigation, and conservation, in addition to hydropower. High-head, low-storage dams tend to be higher-elevation dams that have smaller storage capacity and are used primarily for hydropower. In California, the bulk of the state's hydropower generation is from high-head, low-storage dams, whereas most of the state's reservoir capacity is at lower elevations.[7] Because they have less ability to store water, California's higher-elevation dams are also more vulnerable to changes in runoff.

Climate change affects hydropower availability through changes in the level and timing of temperature and precipitation. Rising temperatures increase the amount of precipitation that falls as rain rather than snow in the winter, and, by reducing the size and extent of the Sierra snowpack, lead to earlier and reduced summer runoff from snowmelt. Changes in the amount of precipitation have a more direct effect, either increasing or decreasing river flows and the annual volume of water available for power generation. The expected effects of climate change on total annual precipitation remain uncertain.

In addition to quantity, the timing of changes in hydropower availability is important. Physical laws dictate that the supply of and demand for electricity must balance at all times on the grid. Presently, electricity cannot cost-effectively be stored on a large scale, which means that "peaker" power plants must often be built to run for only a few hundred hours per year during summer peak demand, which significantly raises the cost of providing electricity during these hours. In contrast, hydropower plants can store water in their reservoirs, giving them the ability to vary output based on demand. This makes hydropower an important resource in hot summers, when afternoon demand is high. Reductions in summer hydropower capacity would increase the need for alternatives in peak periods, which, because hydropower tends to be cheap, would raise system costs.

The ability to control water releases from their reservoirs gives hydropower plants considerable flexibility in when they generate electricity. This flexibility is important for maintaining balance between electricity supply and demand and will become more valuable with higher market penetration by intermittent renewable generation, such as wind and solar. Hydropower is a near-zero CO_2 emission source of electricity, whereas the main alternative form of flexible generation, natural gas, has a CO_2 emission factor that ranges from roughly half of to only slightly less than that of coal, depending on power plant efficiency and the method of extracting the gas.[8] Loss of hydropower capacity would mean either using more natural gas for backup generation, which would be expensive under a carbon pricing scheme, or significantly expanding the scale of other forms of flexible generation, such as electricity storage.

A number of studies have estimated changes in hydropower generation capacity in California and the Pacific Northwest under different climate warming scenarios (table 9). Qualitative differences between the regions are misleading, as the Pacific Northwest results are based on projections of wetter autumns and drier summers in that region,[9] whereas the California studies intentionally used wet and dry scenarios. In reality, both regions face similar hydropower challenges, as rivers in both regions are heavily reliant on snowmelt.

Qualitatively, the studies cited in table 9 offer three key

TABLE 9

Projected Changes in Annual Hydropower Generation with Climate Change in California and the Pacific Northwest

Region	Scope	Climate model–emissions scenario	Change in annual generation	Timing of impact	Source
California	137 of 156 high-altitude dams	GFDL-A2 (dry)	−20%	No end year given	Madani and Lund
		PCM-A2 (wet)	+6%		
		Warming only (no change in precipitation)	−1%		
	Upper American River Project	GFDL-A2	−13%	2099	Vicuña et al.
		GFDL-B1	−10%		
		PCM-A2	14%		
		PCM-B1	9%		
	State Water Project	Based on 12 scenarios from 6 general circulation models	(−5%)–(−12%)	2050	Chung et al.
			(−15%)–(−16%)	2099	
	Central Valley Project		(−4%)–(−11%)	2050	
			(−12%)–(−13%)	2099	

Pacific Northwest	Columbia River Basin	Average of 20 general circulation models, B1 and A1B emission scenarios	Avg: (−0.8%)–(−3.4%) Winter: +1.0%–4.5% Summer: (−8.6%)–(−11.0%)	2020s	Hamlet et al.
			Avg: (−2.0%)–(−3.4%) Winter: +4.7%–5.0% Summer: (−12.1%)–(−15.4%)	2040s	
			Avg: (−2.6%)–(−3.2%) Winter: +7.7%–10.9% Summer: (−17.1%)–(−20.8%)	2080s	
		4 emissions scenarios, 4 climate models, 2 models of the Columbia River Basin	(−16%)–(+3%)	2020s	Markoff and Cullen
			(−30%)–(+2%)	2050s	

NOTE: GFDL and PCM are low-sensitivity climate models. Climate sensitivity is the responsiveness of the earth's near-surface air temperature to changes in radiative forcing. B1 is a low greenhouse gas emissions scenario; A2 is a business-as-usual emissions scenario; A1B is a medium-high emissions scenario. Where reported by the authors, precipitation scenarios are shown in parentheses.

SOURCES: Kaveh Madani and Jay R. Lund, "Estimated Impacts of Climate Warming on California's High-Elevation Hydropower," *Climatic Change* 102 (2009): 521–38; S. Vicuna et al., "Climate Change Impacts on High Elevation Hydropower Generation in California's Sierra Nevada: A Case Study in the Upper American River," *Climatic Change* 87 (2008): S123–S137; Francis Chung et al., "Using Future Climate Projections to Support Water Resources Decision Making in California," California Climate Change Center Report, CEC-500-2009-052-F, 2009; Alan F. Hamlet et al., "Effects of Projected Climate Change on Energy Supply and Demand in the Pacific Northwest and Washington State," *Climatic Change* 102 (2010): 103–28; Matthew S. Markoff and Alison C. Cullen, "Impact of Climate Change on Pacific Northwest Hydropower," *Climatic Change* 87 (2008): 451–69.

insights. First, changes in precipitation rather than shifts in the timing of runoff appear to be the larger determinant of future hydropower availability for California.[10] Storage capacity is sufficient to mitigate changes in the timing of runoff for many hydropower facilities. Second, increases in winter precipitation in California result in a relatively limited increase in hydropower generation. Much of the increased runoff is "spilled," released without passing through dam's turbines, because the additional flows exceed the design capacity of turbines or seasonal demand. Third, reductions in physical hydropower availability in the Pacific Northwest may understate the region's reduction in hydropower exports to California over the longer term. The Pacific Northwest currently has a winter peak in electricity demand, but rising temperatures could shift this peak to the summer, reducing hydropower exports to California over and above any reduction from decreased precipitation.[11]

Table 9 illustrates the large degree of uncertainty surrounding climate change impacts on hydropower generation. In a best-case scenario, the impact on hydropower availability in California would be negligible or even positive. A worst case, alternatively, might result in a reduction of 25 percent in annual hydropower deliveries in California by the end of the century.[12] Replacing this generation would lead to an increase in electricity system costs on the order of $1.7 billion per year,[13] which would be 5 percent or less of total system costs.[14] This estimate does not account for carbon constraints, which could greatly escalate the challenge and likely the costs of replacing hydropower.

It is worth noting that most of California's dams are relatively old, with a capacity-weighted average age of forty-seven years in 2010,[15] and that new large dams are unlikely to be built. Though dams may last for more than a century, it remains unclear whether climate change–induced changes in runoff or dam retirement will do more to reduce California hydropower generation by the end of this century.

Increased Summer Electricity Demand

Rising average temperatures and more frequent days with extreme temperatures could affect energy use by increasing demand for cooling and reducing demand for space and water heating. In California, changes in cooling and heating call upon different final energy sources, as most of the state's cooling is done with electric air conditioners and most of the state's space and water heating is done with natural gas furnaces. Air-conditioning is a particularly important part of state electricity demand, accounting for around 30 percent of peak load.[16]

California's topography influences energy use patterns in the state, with sixteen distinct building climate zones that are segmented in large part by terrain. More generally, electricity use in coastal areas tends to be lower than in inland valleys, where higher temperatures induce more demand for air-conditioning. Changes in climate within these two macroregions over the last four decades are consistent with this pattern. Temperature data from 1970 to 2005 indicate that coastal areas have cooled by 0.01°C per decade, due to increased sea breeze, while inland areas have warmed by 0.24°C per decade.[17] Similarly, per capita electricity use in coastal areas appears to be declining, while in inland areas it is rising.[18]

Future electricity demand in California is thus sensitive to both the quantity and the spatial distribution of population growth. The U.S. Census Bureau projects that nearly 50 percent of California's population growth between 2010 and 2050

will be in the San Joaquin Valley and inland Southern California, regions that accounted for just over 20 percent of the state's
population in 2010.[19] In nine of the eleven counties in these two
regions, per capita electricity consumption is above the state
average.[20]

Studies of climate change impacts on California energy use
date back to the early 1990s, before the development of more
sophisticated climate models (table 10). Improvements in climate
models and higher resolution data have allowed more recent
analyses to account for a range of climate scenarios, as well
as nonlinearities and spatial diversity in how electricity users
respond to changes in temperatures.

All of the studies in table 10 conclude that rising temperatures will drive increases in electricity use, by some estimates
substantially. In a worst-case scenario, based on the estimates
by Auffhammer and Aroonruengsawat, an increase in electricity demand in both the residential and commercial sectors of
55 percent over 1980–2000 average levels would mean an additional 80 billion kilowatt-hours (kWh) of demand by the end of
the century, increasing costs by at least $11 billion per year.[21] As
Auffhammer and Aroonruengsawat demonstrate, however, population growth is a much larger source of uncertainty for future
residential electricity consumption than climate change.

How higher electricity costs would impact the California
economy remains unclear. Higher costs only affect prices to
the extent that marginal costs are above average costs. For both
hydropower losses and increases in summer electricity demand,
this would seem to be the case, as replacements for hydropower
are likely to have higher unit costs and electricity is most expensive during peak periods in summer afternoons. How much
electricity prices would rise, and the effect of higher prices on

the state's macroeconomy, are difficult to assess, as these depend on how both the supply and demand sides of electricity markets respond. However, even a 50 percent increase in real prices over the course of a half a century, equivalent to 0.8 percent per year and 7.5 percent per decade, may be small enough to be accommodated by real income growth.[22]

ADAPTATION OPTIONS AND CHALLENGES

Energy has a number of public goods characteristics, from reliability to air quality, that are likely to sustain strong public oversight indefinitely. However, the appropriate role for public agency in energy sector climate adaptation is not obvious. Energy companies are very experienced with their sector's systemic uncertainties and already deploy sophisticated tools to manage these risks. For this reason, as well as the perennial opportunity costs and contending claims for public funding, there is a strong argument for allowing private sector energy companies to participate in climate risk management while maintaining the regulatory oversight and planning capacity needed to achieve public goals.

Technological innovation may be able to address many of the challenges of adapting energy systems to a changing climate. The need for a more resilient electric power grid could be met by gradual changes that might include the following: greater balance between centralized and distributed generation, sophisticated markets for managing peak load (demand response), greater regional interconnection, and a smarter grid that allows problems to be more quickly isolated and bridged. Hydropower attrition could be compensated with a more diverse generation mix, and greater variance in hydropower availability could be addressed through improved forecasting techniques. Growing

TABLE 10

Projected Impacts of Climate Change on Electricity Consumption in California

Sector	Temperature change or climate model–emissions scenario	Period	Baseline	Energy impact, in MWh	Peak demand impact, in MW	Source
Buildings, water conveyance	+0.6°C	2010	Base case	+0.6%	+1.8%	Baxter and Calandri
	+1.9°C			+2.6%	+3.7%	
Residential	B1	2000–2050	2000	n/a	+0.3%–5.4%	Miller et al.
	A2/A1Fi				+1.1%–7.7%	
	B1	2000–2100	2000		(−1.0%)–6.2%[a]	
	A2/A1Fi				+1.2%–14.2%[a]	
All sectors	PCM/GFDL-B1	2035–2064	1961–1990	+1.6%–3.8%	+1.6%–5.1%	Franco and Sanstad
		2070–2099		+2.9%–5.3%	+4.2%–7.5%	
	PCM/GFDL-A2	2035–2064		+2.3%–4.6%	+2.2%–5.2%	
		2070–2099		+4.8%–9.9%	+5.7%–12.4%	
	Hadley3- A1Fi	2035–2064		+8.1%	+11.2%	
		2070–2099		+17.8%	+19.8%	

Residential			1980–2000		n/a	Auffhammer and Aroonruengsawat
	B1	2059		+11%		
	A2			+17%		
	B1	2099		+21%		
	A2			+55%		

NOTE: Hadley3 is a medium-sensitivity climate model; PCM and GFDL are low-sensitivity climate models. B1 is a low emissions scenario; A2 is a business-as-usual emissions scenario; A1Fi is a high emissions scenario.

SOURCES: Lester W. Baxter and Ken Calandri, "Global Warming and Electricity Demand: A Study of California," *Energy Policy* 20 (1992): 33–44; Norman L. Miller et al., "Climate, Extreme Heat, and Electricity Demand in California," *Journal of Applied Meteorology and Climatology* 47 (2008): 1834–44; Guido Franco and Alan H. Sanstad, "Climate Change and Electricity Demand in California," *Climatic Change* 87 (2008): S139–S151; Maximilian Auffhammer and Anin Aroonruengsawat, "Uncertainty over Population, Prices or Climate? Identifying the Drivers of California's Future Residential Electricity Demand," Energy Institute at Haas Working Paper 208, 2010.

a These estimates are based on the Miller et al. results, which are inconsistent.

air-conditioning use and peak electric load could be addressed with more on-peak solar generation, energy storage, and/or a more geographically extensive transmission network.

Regulatory agencies are often encouraged to be agnostic about technology choice, yet there is a clear role for regulators to identify technologies that will facilitate adaptation to climate change and its attendant costs for society. This exercise of public interest requires not only comparing established technologies but also recognizing those that can emerge over the time frame in which they might be needed, and assessing barriers to their development and deployment. This kind of mapping exercise can provide a useful guide to public and private sector research, development, and deployment efforts, which often require significant lead times and a stable vision of the future. California has established processes in place to oversee the pathways for developing and deploying energy technologies that reduce greenhouse gas emissions.[23] The mandate for this kind of strategic visioning processes could readily be expanded to include adaptation technologies.

In the context of energy services, government agencies will also find themselves in the role of adjudicators regarding distribution disputes that can be expected to arise from climate change. For instance, within Pacific Gas & Electric's (PG&E's) service territory, temperature increases are likely to escalate air conditioner use by residents in the San Joaquin Valley. Because electricity billing is stratified with inclining block prices, without more regional differentiation in prices, customers in these areas will shoulder an ever-greater share of electricity costs, effectively subsidizing customers further north. Regulators will need to determine fair solutions to disputes of this kind.

From a historical perspective, it appears likely that most of the energy sector's climate change investment risk will be assumed by private producers. However, because the costs of "used and useful" assets can be passed on to ratepayers, there is a possibility that utilities might unfairly transfer investment and planning risk to consumers. For instance, because of poor planning, utilities might be forced to buy power from expensive power plants to meet demand. In this event, should investors or consumers be required to shoulder the additional cost? Addressing such questions requires clear rules and congruent expectations between regulators and the utilities about who is responsible for managing climate-related energy risks and to what extent.

Over the longer term, increased variability in supply and demand resulting from climate change will pose a challenge for today's consensus on the appropriate industrial organization of California's electricity sector. The regulated utility model that has been prevalent in the United States for the past century has been effective at responding to medium- to long-term policy drivers and achieving public goals, but this model lacks flexibility, is inertial, and inclines toward centralized, capital-intensive solutions. It remains an open question whether the current regulatory model is the most effective approach to fostering a resilient electricity sector that cost-effectively manages variability in supply and demand.

Summing up, the electric power sector comprises less than 1 percent of California's gross state product, but the services of this commodity are so pervasive that they are linked to all economic activity and employment. Of the diverse impacts that climate change may have on California's energy systems, two are most critical. First, changes in the seasonal availability of water would lead to a reduction in the state's hydropower resources, which could fall by as much as 25 percent from current levels by the end of the century. Second, an increase in the number of extreme heat days could lead to a substantial rise in electricity demand for use in air-conditioning. Both of these impacts can be tempered by both mitigation and adaptation measures, but they will require proactive strategies from a variety of state agencies and the private sector.

Both climate mitigation (AB 32) and adaptation imperatives suggest the need for a radical rethinking of electricity production and distribution in California. Renewable energy can make an important contribution in both contexts, particularly distributed technologies such as solar photovoltaics. Demand-side management technologies and policies can also play a dual role in both mitigation and adaptation. A longer-term road-mapping exercise for energy technologies that meet both mitigation and adaptation goals could provide guidance that supports technology innovation and adoption.

CHAPTER FIVE

Transportation

Transportation infrastructure provides essential supporting services for California's economy and society. Within the state, air, rail, water, and truck transportation accounts for more than $20 billion of California's gross state product.[1] Trade, a core part of the state's economy, has grown in importance along with Asia's economic rise over the last several decades, as California's ports and freight infrastructure provide a convenient gateway for U.S. exports and imports.[2] California is also a highly automotive society, dependent on roads, bridges, and other publicly supported transportation services.

Climate and weather have disproportionate impacts on transportation systems. Landslides damage roads and disrupt traffic. Inclement weather shuts down airports. Extreme heat buckles pavement, requiring road closures and repairs. Climate change has the potential to make all of these problems worse while adding new ones. Sea level rise, for instance, could inundate coastal infrastructure and amplify the adverse effects of Pacific storms. Because of the network characteristics of transportation, individual mishaps have multiplier effects, and adapting the transportation system to changes in climate is a necessity rather than a choice.

Transportation infrastructure tends to be long-lived, more so than in other sectors examined in this study. As a result, proactive and longer-term transportation planning—across a diverse array of federal, state, regional, and local agencies and the private sector—will be the key to effective and efficient adaptation to an uncertain future climate.

ECONOMIC IMPACTS OF CLIMATE CHANGE

To a greater extent than in other sectors, the economic impacts of climate change on the transportation sector are closely tied to the continual cycles of infrastructural retirement and renewal. Most transportation infrastructure decisions play out over many decades (see table 11), and they often extend long beyond intended design lifetimes. Thus new infrastructure commitments are usually made with substantial uncertainty about future operating conditions, but climate change will greatly increase this uncertainty. To adapt it to a changing climate, existing infrastructure may require significant and expensive upgrades (such as the elevation of large sections of coastal highways, for example), while other facilities may have to be prematurely retired.

Climate change will have diverse impacts on California's surface, water, air, and marine transportation infrastructure (see table 12), but the two main challenges it faces are the interrelated effects of sea level rise and storm activity.

As we discuss in more detail in chapter 7, there is at present a wide range of projections for sea level rise (fig. 7). By the

TABLE II

Design Lifetimes of Selected Transportation Infrastructure

Mode	Infrastructure	Design lifetime
Surface transportation	Pavement	10–20 years
	Bridges	50–100 years
	Culverts	30–45 years
	Tunnels	50–100 years
	Railroad tracks	50 years
Aviation	Runway pavement	10 years
	Terminals	40–50 years
Marine	Docks and port terminals	40–50 years
Pipelines	Oil and gas pipelines	100 years

SOURCE: Estimates are based on National Research Council (NRC), "Potential Impacts of Climate Change on U.S. Transportation," Transportation Research Board Special Report 290, 2008.

end of the century, the highest of these, at nearly six feet (1.8 meters) above current averages, could cause major disruption to California's water, ground, and air transportation systems. The state's major ports (Long Beach, Los Angeles, and Oakland) are the facilities most obviously vulnerable to significant rises in sea level. A small but important part of California's rail and road network, including the Pacific Coast Highway, will also be directly vulnerable. California also has two major airports, San Francisco and Oakland, at which most of the runways are less than a meter above sea level. Given the enormous use value of this infrastructure, it will have to be protected, moved, or rebuilt elsewhere, at a cost that could range from reasonable to exorbitant, depending on the timing and effectiveness of longer-term planning efforts.

Pacific storms already cause havoc across California's trans-

TABLE 12

Potential Climate Change Impacts on California's Transportation Infrastructure

Climate impact	Physical impact	Infrastructure
Sea level rise	Inundation	Ports Roads Railroads Airports
Increase in frequency and intensity of major storms	Landslides Flooding High winds Intense waves Storm surge Accelerated coastal erosion	Ports Roads Bridges Railroads Airports
Increase in frequency and intensity of heat waves	Pavement stress	Roads Airports
Rising temperatures	Change in construction material durability	Ports Roads Bridges Railroads Airports

portation system, a hazard that is likely to be exacerbated by climate change. Of particular concern is the El Niño–Southern Oscillation (ENSO), a cyclical weather phenomenon that occurs every two to seven years, driven by the warming of Pacific surface waters. ENSO's effect on California is erratic, but in strong ENSO years, such as 1982–83 and 1997–98, storm damage can be considerable. In other years, ENSO events result in drier weather.[3] ENSO patterns are very difficult to model, and there is still disagreement about how changes in climate would

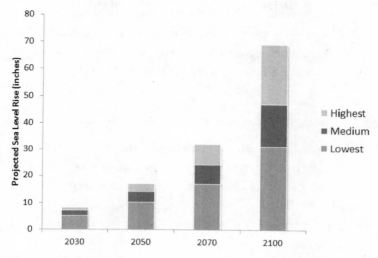

Figure 7. California Department of Transportation working estimates of sea level rise. Source: California Department of Transportation, "Guidance on Incorporating Sea Level Rise," Document for Use in the Planning and Development of Project Initiation Documents, May 16, 2011.

affect ENSO patterns.[4] The costs of ENSO to California's transportation infrastructure are also difficult to assess, but they are thought to be hundreds of millions of dollars per event.[5]

In addition to ENSO events, California's transportation system faces an array of storm risks, from more intense storm surge and wave action that would be aggravated by sea level rise to megastorms, both discussed in more detail in chapter 7. Potential damages to the transportation sector from such storm risks are difficult to assess precisely, and few attempts have been made to do so. Porter et al. estimate that, if there were a megastorm, the cost of repairing road damage from landslides could rise into the billions of dollars, a significant increase over current costs, but tractable given that megastorms are still seen as remotely likely, worst-case scenarios.[6] A potentially larger cost to the transportation network would be prolonged service interruptions at key nodes, highlighting the need for engineering and building in resiliency and redundancy as an adaptive strategy.

ADAPTATION OPTIONS AND CHALLENGES

Both public and private observers agree that California's existing transportation system is far from optimal. Much of the state's transportation system is in a state of serious disrepair[7] and is rooted in fossil fuel use patterns that are outdated, requiring significant new investment to make it both more functional and more sustainable. In its "California Infrastructure Report Card 2006," the American Society of Civil Engineers estimates that bringing the state's aviation, ports, and surface transportation infrastructure up to a "B" grade would require additional annual investments of $20 billion,[8] which would more than double the total amount that the state currently spends on transportation.[9] In addition to allowing a fundamental reconfiguration of the transportation system, the need for significant investment presents an opportunity to make the transportation system more resilient to climate change.

The cost of transport sector adaptation depends upon the effectiveness of planning efforts. Proactive long-term planning can steer new infrastructure toward less risky areas, impose design standards that accommodate changing climate conditions, ensure that older infrastructure is either protected or phased out before it sustains significant damage, and build more resilience and redundancy into the system to allow it to recover from damage more quickly. Alternatively, reactive planning will require extensive emergency engineering efforts and contingency resources, which are usually very expensive. For instance, if sea levels were to rise rapidly after 2050, a planning process begun decades earlier could have identified solutions for protecting or rebuilding airports. A reactive approach would force decision makers to scramble for solutions, such as raising airports or moving them offshore. As a reference point for the costs of such an approach, building Hong Kong's Chek Lap Kok Airport on landfill in the South China Sea in 1998 cost $20 billion.

Increases in storm intensity can be better anticipated with forward-looking design, which comes at a cost but can be done cost-effectively if the potential damage justifies the investment. The New Zealand government, for instance, estimates that the cost of redesigning and retrofitting its bridges for climate change raised initial costs by 10 percent, but over the life of the structures the additional cost amounted to only 1 percent because of damage that was avoided.[10] The difficulty is determining what damages might be expected, which may be as much a question of risk tolerance as it is a question of science.

"Softer" adaptation solutions, such as optimizing capacity usage with intelligent transportation systems, could be a cost-effective means of making the transportation system more resilient in the coming decades. Because of the high cost of land, transportation systems are generally not flexible spatially in the near to medium term, which makes expanding the system prohibitively expensive. In the longer term, strategically extending both transportation modes and infrastructure over several decades could provide greater resiliency and positively influence development patterns, but this, too, requires proactive longer-term planning. The centrality of long-term planning across a range of adaptation solutions highlights the need for strong, capable public agencies and planning processes.

Integrating climate change adaptation into transportation planning in California could be a complex and multilayered affair, as transportation decisions are made by a multitude of entities, both public and private, from the federal level down to local jurisdictions, which have different interests and planning horizons. The California Department of Transportation (Caltrans) manages statewide surface transportation planning. On a parallel plane, eighteen Metropolitan Planning Organizations (MPOs) and forty-eight Regional Transportation Planning Agencies (RTPAs) work with local governments on planning and programming at the regional and local levels. The actual development and operation of the transportation system is the responsibility of a combination of different levels of government along with private agencies.

Caltrans typically owns and operates a relatively small portion of the road network, but that portion, the interstate system and arterial highways, usually accommodates the bulk of road travel within the state. The majority of highway miles and transit systems are the responsibility of the regional and local governments. Caltrans decisions encompass all aspects of the roadways under state jurisdiction, including planning, engineering, operations, design, and construction. Caltrans's California Transportation Plan, a statewide, long-range transportation plan with a minimum twenty-year planning horizon, is prepared in response to federal and state requirements and updated every five years. At the MPO level, decisions represent a collaboration of the individual local governments that comprise the MPO and serve on the policy boards supported by the advice and analysis of a technical coordinating committee. The Regional Transportation Plans at the MPO/RTPA level are conducted on a twenty-five- to thirty-year time horizon.

California's freight system and airport/passenger system are owned by the private sector. Although Caltrans and the MPOs with local jurisdiction may have some influence through the planning process or through the provision of financial support, they must obtain agreement from the private sector for implementation. Typically, the private sector invests in their current system or a new one only if they feel it is cost-effective to do so. For instance, a railroad will not likely move a rail line unless it improves their return on investment or unless the government helps finance it. For California's transportation system to effectively and efficiently move people and goods, as well as meet the particular challenges that climate change poses, a high degree of public and private sector coordination is essential. The different levels of government and private entities will have to understand how climate stressors might affect their own assets, as well as one another's, in order to take full advantage of system redundancy and resilience.

These institutional dynamics raise another important question: Who should pay to improve the adaptive capacity of the transportation system? In addition to road infrastructure, which is financed primarily by the public, significant state, local, and often federal funds are committed to airports, railroads, and marine ports when improvement programs are undertaken. Should state and local governments shoulder the incremental cost of adaptation-related investments in the transportation system? Should users pay this cost through user fees? How these investments are paid for and priced will influence the effectiveness of the transportation system's adaptation to climate change, as well as the overall evolution of California's transportation system.

Transportation is the foundation of the state's spatially diverse but highly integrated economy, as well as its lifeline to the national and global economy. Its ports account for a large share of U.S. container volume and total foreign trade. Climate change will be an important consideration for transportation planning in California. However, the vulnerability of California and U.S. transportation infrastructure to climate change and the extent to which it may be avoidable, are still poorly understood. Only a small amount of research has been done on transportation sector climate adaptation.

Climate impacts on the state's transport infrastructure are likely to be concentrated in coastal areas, where sea level rise and storm and wave action will threaten maritime facilities, airports, and coastal/fluvial ground transport assets. In the Bay Area, for example, all three major airports are near sea level. They could be raised or barricaded against sea level rises, but storm and wave action might pose unacceptable risks. Because of easement issues in a dense metropolitan area, the cost difference between fortification and relocation could be one or even two orders of magnitude. Expected costs of maritime and ground transport defense are currently hundreds of millions per year. These numbers are within range of current public infrastructure budgets but remain quite speculative. Among other things, these estimates include annualized costs of intermittent disasters, as well as significant uncertainty about the amplitude and frequency of Pacific storm and tidal cycles. Improving these estimates should be a high priority, however, because the use value of these assets is enormous and their lifespan very long, and thus the timing and sequencing of adapting investments is critical.

In light of expected population growth, less intensive use would not appear to be an option for adapting the state's transport infrastructure. Although there has been virtually no research in this area, the remaining options should certainly include more stringent design standards for new structures and retrofits to existing structures. Both would likely require some increase in costs, but engineering solutions of this kind are in most cases likely to be cheaper than relocation. In any case, foresight will again save money by avoiding irreversible misallocation and parallel capacity losses from extreme events, but early action must be better informed.

Soft adaptation options could also be important for reducing climate damages, both inside the transport sector and more broadly. Transportation systems shape land use, population growth, and economic development patterns, and more adaptive transportation planning decisions could have a significant influence on property-related climate damages. Policies that discourage settlement in high-risk areas, for example, will reduce the state's long-run adaptation costs. In the truest sense, these linkages reveal the extent to which climate issues are lifestyle issues.

CHAPTER SIX

Tourism and Recreation

With the state's celebrated combination of beaches, vineyards, winter sports, national parks, and theme parks, tourism from all over the world is an important contributor to the California economy. The California Travel and Tourism Commission estimates that direct travel spending in California was $96 billion in 2010.[1] A significant portion of tourism in the state is undertaken by California residents, and we add recreation to the title of this chapter to emphasize that fact.

Many of California's top travel destinations are dependent on natural resources, including its beaches, rivers, mountains, vineyards, golf courses, and national and state parks. Climate change is expected to affect all these venues, but in different ways. In lower warming scenarios, some tourist/recreation sites and activities would experience negative net impacts while others would benefit. For instance, some warming may be good for the golf industry, whereas any warming is likely to have adverse effects on winter sports. For the United States as a whole, earlier studies argued that climate change would, on balance, benefit the tourism sector because summer recreation industries are larger economically than winter recreation.[2]

Regardless of the ultimate economic impact, the tourism sector will face important adaptation challenges in California because of its reliance on the natural environment, which will be transformed through changes in climate. How much should be done and spent to try to maintain the current natural environment versus letting it adapt is an important decision process that, although only in its beginning stages, has far-reaching societal and political consequences. The U.S. Congress, for instance, has a mandate to protect the national parks for future generations,[3] but it has yet to determine precisely what that this means for the current and future use of the vast expanse of federal land in forest and coastal areas.

ECONOMIC IMPACTS
OF CLIMATE CHANGE

Climate change is expected to have diverse economic impacts on California's tourism and recreation industry, including those sectors presented in table 13. Although this table is by no means exhaustive, three leading industries—beaches, winter sports, and outdoor recreation—have been the most intensively studied and are the focus of our discussion here.

Beaches

Sandy beaches, inherently unstable, are one of California's most lucrative but vulnerable natural resources. Sand is deposited on beaches from river sediment, eroding bluffs, and migrating sand from other beaches. From there, it is removed or inundated by a

TABLE 13

Potential Climate Change Impacts on Tourism and Recreation in California

Climate impact	Physical impact	Affected industries
Sea level rise, changes in Pacific storm and tidal activity	Reduction in beach width from inundation and erosion	Coastal water sports and beaches
Rising average temperatures, changes in precipitation	Shortening of the snow season Loss of glaciers and snowfields Changes in vegetation Wildlife population and habitat change Park closures due to fire and heat	Winter sports industry National and state parks
Changes in stream flows and temperatures	Changes in aquatic and riparian habitats	Recreational fishing, hunting, and aquatic and allied outdoor sports
Rising average temperatures, changes in weather variability	Increases in average and variance of number of fair-weather days	Golf, sailing, and other fair-weather outdoor sports

host of spontaneous natural processes that include wave activity and sea level changes, the latter due to cyclical tidal forces or global temperature regimes. Over the course of the twentieth century, the building of dams, reservoirs, and, to a lesser extent, coastal infrastructure has significantly reduced the volume of sediment reaching the California coastline as sand, abetting beach erosion.[4] To counter these trends, Californians have restored beaches and protected them from further erosion by adding sand dredged or trucked in from elsewhere, a practice known as "beach nourishment."

Of all the activities in California's tourism portfolio, beach recreation has the highest economic exposure to climate change. Estimates of the value of the state's beaches vary but are on the order of tens of billions of dollars per year in revenues and billions of dollars per year in value added.[5] By reducing their width, inundation and beach erosion could decrease the number of people visiting beaches, and, as a result, adversely impact local economies that are dependent on beach-related spending. Although the reduction in beach width could be substantial—reaching an average of sixty-two feet (nineteen meters) in Southern California by the end of the century—these reductions are likely to have only a marginal impact on beach going, given the large number of wide beaches in California.[6]

Pendleton et al. estimate that a one-meter rise in sea levels and an increase in storm activity would lead to a small net reduction in Southern California beach visits and revenues by the end of the century,[7] but the consequences across beaches would be very uneven. A small net change in beach going can mask significant diversion of patronage because beaches have different adaptation capacity. Large beaches, where the reductions in beach width would be less noticeable, would see more traffic, where smaller beaches, for the opposite reason, would see less. At the extremes, a one-meter rise in sea level is projected to increase spending at Huntington Beach by $16 million per year, while it would decrease spending at Laguna Beach by $14 million per year.[8] Alternatively, it is projected that an extreme storm year would increase spending at Laguna Beach by $20 million and reduce it at Redondo Beach by $25 million.[9]

Winter Recreation

Significant long-term reduction in the Sierra snowpack would have profound implications for California's winter sports industry. Ski resorts require a minimum snow depth to begin the ski season, with a typical ski season in California lasting from late November to June, or around two hundred days.[10] Without snowmaking, climate change will shorten the ski season considerably by the end of the century, with projected reductions ranging from forty-nine days to the entire season, but nearer-term impacts are less certain.[11] Shortening of the ski season due to climate change happens at both ends; the onset of the ski season is delayed by the fact that temperatures are not cold enough, while the season ends sooner because of earlier snowmelt.

Because lower-altitude ski resorts are likely to be more vulnerable to rising temperatures, climate change would likely push California's winter sports industry to higher altitudes, leading to higher prices as supply is reduced. The economic impact on the industry is uncertain. The California Ski Industry Association estimates that winter sports in California generate $500 million in annual spending.[12] If the entire ski season were to disappear by the end of the century, as in Hayhoe et al.'s most extreme scenario, this entire $500 million would be lost. Even in more benign scenarios, losses to the winter sports industry could be in the hundreds of millions of dollars, depending on how much rising temperatures shorten the season, the cost-effectiveness of snowmaking, and how much the industry is able to raise prices.[13]

Outdoor Recreation

From parks to fishing, nature recreation is an important part of the California economy. According to Fish and Wildlife Service surveys, 8.3 million people fished, hunted, or watched wildlife in California in 2006, spending $7.4 billion.[14] State and national parks account for a large share of this economic activity. The National Parks Conservation Association estimates visitors to California's twenty-three national park sites spent a total of $1.2 billion in 2001.[15]

California's state and national parks cover an exceptionally diverse range of ecosystems, and climate change will have different impacts across parks because of this diversity. Some of these impacts could be devastating, others benign or even positive. Although there is no evidence to suggest that climate change would, on average, reduce park attendance, climate change will likely raise the costs of maintaining the ecological integrity of park systems. State and national parks may have to purchase land to protect habitat corridors as they change, for instance, which would be expensive because of California's high land costs.[16] Escalating fire risks will also raise operating costs and lower the recreational value of these lands.

Recreational fishing is also an important part of California's tourism industry, generating $2.4 billion in expenditures from 1.7 million anglers in 2006.[17] Changes in stream flow and temperatures that result from climate change could have a deleterious impact on fish habitats. In 2002, for instance, lower-than-average water levels and higher temperatures in the Lower Klamath River may have contributed to the disease that caused a massive die-off of more than 33,000 salmon and steelhead trout.[18] In higher warming scenarios, higher water temperatures will threaten the survival of the chinook salmon in the Central Valley.[19] The economic impact on recreational fishing will depend on water managers' ability to maintain fish habitats, which, as we described in chapter 3, will require fundamental changes in institutions and practices related to water allocation and reservoir management.

Climate change could potentially have a positive impact on golf and other fair-weather sports (e.g., sailing and cycling) due to an increase in the number of days in the year conducive to these activities.[20] However, although the number of fair-weather days might increase, an increase in weather variability might offset these benefits. Additionally, in water-stressed regions, golf, at least in its current form, could become untenable because of the inability to maintain greens and fairways. Climate change is already having an impact on the golf industry across the United States. For instance, a study of weather patterns from 1977 to 2006 by WeatherBill found that while the average number of golf playable days (GPD)[21] in the United States—and for five of the eight cities examined in California—has increased over the past three decades, so has GPD variability.[22] This added uncertainty makes planning more difficult for golf course owners.

ADAPTATION OPTIONS AND CHALLENGES

Although studies thus far suggest that, apart from the winter sports industry, climate change could have a relatively limited overall economic impact on California's tourism and recreation sector, a number of important adaptation questions still face the sector.

Tourism and recreation tend to support localized economies and, as one might expect, the impacts of climate change will be uneven across counties and communities. Reductions in beach width, for instance, could be a boon to some cities at the expense of others. Shrinking of the snow season could disproportionately affect lower-altitude resorts and resort towns. How local governments and communities can respond to the gradual—or potentially rapid—disintegration of bedrock industries and whether there should be a state or federal role in supporting those communities are important open questions.

To some extent, climate change impacts can be moderated by preventative measures. Beach nourishment, for instance, can offset beach erosion. Snowmaking can extend the ski season at viable resort elevations. More extensive land conservation can reduce ecosystem vulnerability in state and national parks. Changes in reservoir management can provide colder stream water for fish. These activities all come at a cost. Pendleton et al. estimate that protecting beaches from a one-meter inundation would require an average of $4 million per year over the next century, a 40 percent increase over current expenditures, but that a series of severe storms could require nearly $400 million in nourishment in a single year.[23] Snowmaking is also expensive, with a price tag of hundreds of dollars to make just one acre-foot of snow.[24] Conservation through land purchases can cost thousands of dollars an acre.[25]

For all industries, a central question will be who pays the additional cost. Beach nourishment is currently financed by a combination of federal, state, and local government funding.

Pendleton et al. argue that nourishment would be economically justified as long as the cost is less than the total market and non-market value of beaches to Californians, but this calculus avoids the questions of public/private cost sharing, whether state and federal funds are better spent elsewhere, and whether and how much ecosystems should be allowed to adapt.[26] Snowmaking is currently paid for by resort owners, but if beaches receive public funding, why should the ski slopes not? For California's park system, there is a clearer role for federal and state management, but how much of the additional costs should be paid for by taxpayers and how much through increased user fees? These questions, all involving the appropriate collection and use of public funds, need principled solutions, which will require both time and political will to formulate and implement.

Tourism is a major activity and source of income and employment in California, and because the majority of people classified as "tourists" in the state are actually residents, tourism and recreation can be considered as one sector. Many of California's top tourism destinations, including beaches, ski resorts, state and national parks, golf courses, and playing fields, are outdoors. Climate change is expected to affect all of these venues and their appurtenant activities, but in different ways. For some industries, such as golf, lower levels of warming will be beneficial. Others, like the California ski industry, are threatened with extinction. Water sports will likely be attenuated somewhat on an annual basis and shift more strongly toward a seasonal basis. The cumulative cost estimates for these impacts still vary considerably, between hundreds of millions and billions annually. More important, however, may be the distributional impacts, which will shift income and employment around the state and between very different activities.

For this sector, the financial burden of adaptation will be relatively evenly divided between public and private stakeholders, at least if the existing patterns of asset ownership, management, and use are followed. Adaptation options differ for each of the three main categories of recreation (beaches, winter sports, and other outdoor activities). Most of the state's recreational beach capacity can be sustained or substituted with "nourishment" and protection strategies. For winter sports, artificial snowmaking can extend or defend useful facility range and life, but this appears to be part of an end-game process that should include non-winter-use diversification. Other outdoor activities can best adapt with strategies that combine diversification with ecosystem conservation that improves the recreational capacity of public and private lands.

CHAPTER SEVEN

Real Estate and Insurance

For the last two decades, real estate has been at the core of the California economy, nearly 50 percent larger than any other sector and accounting for one-sixth of gross state product (GSP), or $281 billion in 2009.[1] A significant portion of California's building stock is in high-risk areas, in earthquake zones, on the fire-prone wildland-urban interface, in floodplains, and along the coast. Although California has extensive experience dealing with earthquakes, climate change presents a different set of challenges, some familiar and some new, but all serious threats to the state's largest asset class.

Because of its high value, residential and commercial property value will likely be the most important climate risk for California, both economically and politically. The extent of property damages from climate change, and who pays for these damages or adaptation measures to avert them, depends in significant part on how much property is located in high-risk areas decades into the future. In theory, rising risks would encourage property holders to reinvest in less dangerous areas. In practice, these kinds of adjustment processes tend to be slow, politicized, and distorted by policies and institutions. Uncertainties about the actual impacts of climate change only compound the market imperfections that inhibit responsible adjustment to risk.

ECONOMICS IMPACTS OF CLIMATE CHANGE

In California, climate change threatens the real estate sector through property damage stemming from wildfires, sea level rise, coastal erosion, winter storms, and changes in precipitation and hydrology (see table 14).

There is, as yet, nothing resembling consensus on how a warming climate might affect winter storms and flooding. Modeling work thus far suggests that changes in both are likely to be relatively small. Downscaling results from global climate models shows little change in the average frequency of large storms in California over the next century,[2] although they may become more clustered, with multiple storms occurring in a short period of time and then a longer interval before the next storm cluster.[3] This kind of clustering can overtax capacity for both emergency response and recovery. Preliminary hydrological modeling of warming impacts on California's flood regime shows a moderate increase in the frequency and severity of floods.[4] The other two leading threats, wildfires and rising sea level, have been more extensively studied, and their physical impacts are better understood.

Wildfires

Much of California's fire risk is concentrated in the foothills of the Sierra Nevada mountains and along the state's southern coast.

TABLE 14

Potential Climate Change Impact on California's Real Estate

Physical change	Source of property risk
Reduction in vegetation moisture from increased evapotranspiration and reduced summer water availability	Increase in frequency, intensity, and scale of wildfires
Longer fire seasons resulting from reduced summer water availability	
More lightning strikes resulting from a drier atmosphere	
Changes in vegetation moisture	
Sea level rise	Permanent inundation
	Increase in intensity of storm surge
	Accelerated coastal erosion
Increase in frequency, duration, and intensity of Pacific storms	Increase in flood and strong wind probability
Earlier snowmelt	Increase in flood probability
Increased share of winter precipitation falling as rain	

These two regions have different kinds of fire regimes. The wetter and cooler forests of Northern California and the Central Valley have an "energy-limited" fire regime, meaning that fires tend to be limited by the flammability of vegetation. In energy-limited regimes, drier conditions and warmer temperatures increase the risk of fire. The drier and hotter grass and scrublands of Southern California have a " moisture-limited" fire regime, meaning that fires are limited by the availability of vegetation (i.e., fuel) that could be burned. In moisture-limited regimes, periods of high rainfall increase vegetative growth, providing fodder for fires once the weather returns to drier and hotter conditions.

A large portion of California's housing stock lies at the intersection of urban and natural environments, the wildland-urban interface, where brush or forest fires encroach on relatively dense residential real estate. As of 2004, the California Department of Forestry and Fire Protection estimated that about five million homes, or nearly 40 percent of the state's residential housing stock, were located in areas with a fire threat of "High" or greater.[5] At a current average market value of $360,000 per home, these houses represent around $1.8 trillion in asset value.[6] At an estimated average replacement cost of $130 per square foot, rebuilding them would require around $1 trillion.[7] The value of housing in high wildfire risk areas explains the scale of state and federal fire suppression efforts in California, which reportedly exceed $1 billion annually.[8]

Wildfires are closely tied to longer-term climate trends because of their dependence on changes in moisture levels that in turn depend on interactions between temperature and precipitation. Westerling et al. project that, depending on the climate scenario, climate change will increase the frequency of large fires statewide by 28 to 33 percent by midcentury and by 38 to 84 percent by the end of the century.[9] However, these statewide averages mask significant spatial uncertainty. In a separate study, Westerling and Bryant predict that, although the frequency of large fires increases across all climate models in Northern and central California, in moisture-limited southern California, wetter climate models increase the frequency of large fires by around 30 percent, while drier climate models reduce it by around 30 percent.[10]

Actual property damage from future wildfires depends on a number of factors, such as the size and spatial distribution of population, home sizes and values, and protection efforts. Bryant and Westerling estimate that average annual damages to residential property from wildfires could reach billions of dollars per year by midcentury, even in a lower warming scenario, and reach tens of billions of dollars per year by the end of the century.[11] Most of these damages are predicted to occur in the Sierra foothills northeast of Sacramento, the Bay Area, and coastal Southern California.[12]

Direct and Indirect Impacts of Sea Level Rise

As the earth's oceans warm and expand, with rising temperatures melting the earth's stores of ice, climate models predict significant increases in global mean (eustatic) sea levels. For California, earlier analysis suggested that sea level rise might range from 2 to 13 inches (0.06 to 0.32 meters) by 2035–64 and 4 to 28 inches (0.11 to 0.72 meters) by 2070–99, depending on the emissions scenario.[13] More recent analysis projects much larger rises in sea levels along the California coast, between 20 and 55 inches (0.5 and 1.4 meters), again depending on the emissions scenario, by the end of this century.[14]

The real estate risks of sea level rise and coastal erosion are concentrated around California's major bays—Humboldt, Monterey, and San Francisco—and the Southern California coast, a relatively small fraction of the state's 1,100-mile coastline.[15] This contrasts with Gulf Coast states, where most of the coastline is subject to "very high" risk from sea level rise.[16] Thanks to tectonic forces, namely subduction of the Pacific Plate beneath western North America, most of California's coastline is adorned with a picturesque array of high cliffs. These confer significant protection against tides, wave action, and storm surge, making the west coast of the United States much less vulnerable to sea level rise than the east coast, which has extensive low-lying coastal plains.[17] Nevertheless, with roughly 15 percent of its population and an even higher-value share of its housing stock concentrated in coastal areas, climate change has important implications for California's coastal real estate.[18] At a current average market value of $360,000 per home, houses in near-shore areas represent around $1 trillion in asset value.[19] The high value and concentration of near-shore real estate also means that protecting these homes is likely to be more cost-effective than abandoning them, but financing this protection could have important equity implications.[20]

The direct property risks from sea level inundation depend on the longer-term predictability of sea level rise, as well as the frequency and severity of storm surge (discussed below). For example, if sea level rise is both gradual and predictable, the value of coastal property would decline to zero by the time of inundation, with property costs both literally and figuratively sunk. If sea level rise is uncertain, inundation could lead to significant losses because of the high value of near-shore property. For instance, Gleick and Maurer estimate that the value of property at risk from inundation by a one-meter rise in sea levels in the San Francisco Bay alone would be $48 billion.[21]

Coastal erosion and storms are tied to sea level rise, which accelerates erosion and amplifies the effects of storm surge. Higher coastal erosion and more intense storms would in turn increase the damages from sea level rise. Heberger et al. estimate that, by the end of the century, a 1.4-meter rise in sea level along the California coast would lead to $100 billion (in 2008 dollars) in property replacement costs from a one-hundred-year flood, assuming no protection.[22] Damage from coastal erosion, although significant, is predicted to be much smaller than damage from flooding.

Historical Disasters

The past provides useful context for gauging potential property damage from extreme weather events. Table 15 shows California's history of federally declared disasters from 1960 to 2009, indicating events where the extent of damage exceeded the capacity of local resources. Freezes generally affect crops rather than property and earthquakes are not weather related, but the first two columns provide context for the three weather-related property disasters: fires, storms, and floods.

Severe storms and flooding have been the most frequent disasters in California over the past five decades, averaging nine events per decade from 1960 to 2009 and accounting for nearly 70 percent of the state's major disasters over this time period. High frequency does not always imply high damages, however. Of the recent billion-dollar property-related disasters in or immediately involving California, wildfires have been the most frequent, at roughly four per decade from 1990 to 2010. Catastrophic floods have been less frequent but more expensive, at an average of four billion dollars per event (see fig. 8).[23] Interestingly, as we discuss in greater detail in the chapter on agriculture, freezes have been the most expensive weather-related disasters in California over the last two decades, at an average $4.4 billion per event.

Recent history provides only limited guidance regarding climate variability and disaster in California. The largest recorded

TABLE 15

Major Property-Related Weather Disaster Declarations in
California, 1960–2009

	Forest fires	Storms and flooding	Freezes	Earthquakes	Total
2000–2009	3	5	1	2	11
1990–1999	2	8	2	2	14
1980–1989	1	11	0	2	14
1970–1979	1	11	0	2	14
1960–1969	1	10	1	0	12
Total 1960–2009	8	45	4	8	65
Average per decade	1.6	9.0	0.8	1.6	13.0

NOTE: Data are from Federal Emergency Management Agency, "California Disaster History," www.fema.gov/news/disasters_state.fema?id=6#diz.

natural catastrophe to strike California, for instance, was a series of winter storms that caused severe flooding over the course of 1861 and 1862. The floods turned the Sacramento Valley into an inland sea, created lakes in the Mojave Desert and Los Angeles Basin, and, by inundating nearly a third of taxable land, bankrupted the state.[24] An interagency study estimated that a recurrence of a storm of that magnitude in the modern era would lead to property and business interruption damages on the order of $700 billion, nearly three times greater than the expected loss from a devastating earthquake in Southern California.[25] Geologic evidence suggests that California experienced multiple instances of flooding even more severe than the 1861–62 event in the last two millennia.[26]

Given the difficulty of re-creating accurate storm histories over thousands of years, predicting the frequency of these mega-

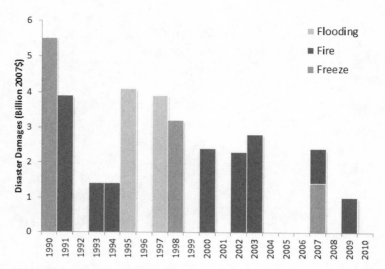

Figure 8. Billion-dollar disasters in or involving California, 1990–2010. Source: National Climatic Data Center, National Oceanic and Atmospheric Administration, "1980–2010 Billion Dollar U.S. Weather Disasters," www.ncdc.noaa.gov/img/reports/billion/disasters2010.pdf.

storms, or how climate change would affect their frequency and severity, remains difficult. However, it is certain that California's current physical and institutional infrastructure do not have the capacity required to deal with disasters of this magnitude.

ADAPTATION OPTIONS AND CHALLENGES

Property damage from weather-related disasters depends on the frequency and magnitude of disaster events, the value per property in high-risk areas, the total amount of property in those areas, and the likelihood of damage. Although the frequency and magnitude of disasters are beyond the control of a state's government and its citizens, damages from disasters are, to some

extent, manageable. Both insurance and mitigation are important tools for managing risk and limiting economic damage.

Insurance does not reduce the total costs of a disaster; it simply spreads risk over time and a larger number of property owners. However, by influencing the risk faced by property owners,
insurance plays a role in determining both where people live
and the value of their property. Insured damages from natural
disasters in the United States have skyrocketed since the early
1990s, in part because of an increase in the frequency of hurricanes.[27] More importantly, however, a lull in weather-related
disasters over the 1970s and '80s led to a significant increase in
the number of people living, and the value of property, in high-
risk areas.[28] Seven hurricanes that struck the United States during 2004 and 2005 inflicted an estimated $90 billion in insured
losses, nearly bankrupting the insurance industry and prompt-
ing a reevaluation of the role of federal and state governments
and the private sector in insuring catastrophic risk.[29]

For California, wildfires and floods are the two main weather-
related sources of catastrophic risk. With a sustained rise in fire
damages over the last decade, the private market for fire insurance has become difficult to sustain in its current form, with
insurers reducing their exposure by canceling existing policies
and denying new ones and by requesting steep increases in premiums from the California Department of Insurance.[30] Because
fire insurance is typically required for a home loan, the availability of coverage has a strong influence on property values.

Since 1968 California has provided limited coverage to homeowners who are unable to obtain insurance in the private market
through the California Fair Access to Insurance Requirements
(FAIR) Plan.[31] Participation in FAIR, though small as a per-
centage of the state's total housing stock, is nonetheless already

significant.[32] An increase in the burned area and intensity of wildfires would test the boundaries of private and public insurance by requiring either a large increase in premiums or, with a reduction in coverage, a greater number of households insured through the FAIR Plan.

Floods have historically been a relatively minor disaster risk for California, and as a result flood insurance coverage is low in scope and cost. Hanak et al. estimate that of the 5 and 13 percent of California households living in 100-year and 500-year floodplains, respectively, only 30 and 7 percent had flood insurance in 2006.[33] The number of businesses in floodplains with flood insurance is unclear.[34] Flood insurance in the 100-year floodplain is provided mainly through the federal government's National Flood Insurance Program (NFIP), which requires homeowners to purchase coverage to be eligible for a mortgage backed by the federal government.

When floods were infrequent and caused limited damage, this arrangement functioned relatively well. However, as more is understood about natural flood cycles, including the fact that flooding might be exacerbated by a warming climate, there are signs that the current system of flood insurance in California is unsustainable. The NFIP is already in debt after a series of expensive disasters in the 2000s,[35] and, more generally, federal disaster insurance may soon face an existential crisis. The federal government's exposure to earthquakes, hurricanes, and floods over the next seventy-five years could be as high as $7 trillion.[36] At the same time, underinsurance in both the 100- and 500-year floodplains poses a major risk to public budgets, as disaster recovery costs would most likely be shouldered by taxpayers.

Creating a balanced, equitable approach to fire and flood

insurance that preserves affordability, incentivizes homeowners and businesses to purchase insurance, encourages policyholders to internalize private risk, and maintains the solvency
of the insurance industry will be a challenge for California's
policymakers. It is unclear, for instance, how insurance companies should price premiums to account for the risks of the megastorms described above, or how premiums should fairly reflect
the risks imposed by climate change, many of which are difficult to predict or even quantify. Even if solutions can be found
for new policyholders, an additional challenge is determining
how to make changes in premiums for existing policyholders
that result from a reassessment of disaster risk gradual enough
to avoid major social dislocation while still minimizing the public sector's exposure. Mitigation measures, by reducing vulnerability, are an essential complement to insurance. This fact
needs to be considered in an environment of rising risk. Publicly
sponsored insurance premiums need to rise accordingly so that
property owners will internalize actual risk more completely
and change their behavior accordingly. This can be politically
difficult, but governments must recognize that subsidizing risk
taking may be very inequitable and in any case depletes public
resources directly and indirectly, increasing the expected cost of
eventual damages and claims.

For wildfire, damage mitigation requires a complete recon-
ceptualization of fire management. Historically, efforts to limit
the damage of large fires in California have stressed evacuation
and fire suppression. This approach does not actually reduce the
risk of property damage, it is expensive, and it leads to large
wealth transfers when the costs of fire suppression are covered
by state and federal taxpayers. Suppressing fires promotes fuel
accumulation and, eventually, more catastrophic fires, while

public financing of firefighting creates moral hazard for property owners in risky areas. An alternative approach recognizes the important ecological role of fires, seeks to find ways to coexist with fire, and emphasizes measures that reduce the risk of damage rather than the frequency of fires.[37] Planning and insurance can play constructive roles in fire mitigation by limiting growth in high-risk areas and encouraging communities and policyholders to undertake proactive measures, such as brush clearing and controlled burning, that reduce the risk of damage.

For floods, mitigation options encompass a spectrum from soft infrastructure, such as flood mapping, zoning, and financial regulation, to hard infrastructure, such as building and maintaining flood defenses and reserving or restoring wetland buffers. Soft options tend to be less capital intensive and more cost-effective, provided that institutions can support them. For instance, because it limits the total amount of damage from flooding, restricting development in floodplains is a particularly effective tool for flood management. However, development limits pit local governments, hungry for tax revenues and with little real or perceived long-term risk from flooding, against state and federal government agencies that may ultimately be liable for damages. Although recent legislation in California has attempted to realign local and state incentives for flood management by forcing local governments to assume some flood liability, the ambiguity of these rules makes it unlikely that local governments will be more cautious about floodplain development.[38]

Hard options, and particularly building and maintaining flood defenses, are typically expensive. Heberger et al. estimate that building seawalls and levees to protect vulnerable areas from a 1-in-100-year flood exacerbated by sea level rise would cost at least $14 billion (in 2000 dollars) and would cost $1.4 billion per year to maintain.[39] Raising funds for flood management infrastructure in California is a perennial challenge. The only funding mechanism the state has for flood infrastructure is bond finance, which does not identify a long-term revenue source and is subject to popular vote. A more sustainable funding source, such as a surcharge on water use, could provide the necessary funds, but it would raise critical questions about who should pay and how much.

In the industrialized world, the economic effects of climate change will be dominated by valuation effects on real property. Whereas in many parts of the developing world other concerns, such as the spread of disease, may take precedence, in California the dominant climate issues will likely be related to demographics, land use, and real estate. Deeply implicated in the same discussion is the insurance industry, whose risk exposure will increase dramatically, offering new opportunities for financial risk management to guide both mitigation and adaptation. Another somewhat perverse benefit of escalating insurance costs will be accelerated adaptation, as an ever-broader array of adaptation investments becomes profitable.

Adaptation strategies for real estate and insurance are very complex from both material and behavioral perspectives. As fixed assets, property values are highly vulnerable to changing local conditions. Residents and values in risk-prone areas should to a significant extent internalize such risks, but if history is any guide they will use political means to resist that. However, the geographic concentration of flood and fire vulnerabilities means that defense and adaptation policies are inherently distributional, using scarce public funds to secure the wealth of some residents but not others. Also, moral hazard in this context may undermine the insurance industry's capacity to price risk accurately, increasing exposure and the ultimate risk of costly public bailouts.

CHAPTER EIGHT

Public Health

Globally, effects on human health may be one of the most visible and provocative consequences of a changing climate. A joint commission from *The Lancet* and University College London referred to climate change as "the biggest global health threat of the 21st century."[1] Although perhaps in less dramatic fashion, climate change will exacerbate existing public health challenges for California, even as it creates new ones.

Californians already have the worst air quality in the United States,[2] with the number of deaths from air pollution on par with those from traffic fatalities and secondhand smoke.[3] The California Air Resources Board estimates that the cost of air pollution to the state is between $36 billion and $136 billion per year, with pollution control programs for industry and motor vehicles costing about $10 billion per year.[4] Climate change has the potential to increase the costs of both pollution itself and pollution control. New challenges, such as more intense heat waves, will put pressure on California's public health systems to increase capacity and be more proactive, targeted, and responsive.

ECONOMIC IMPACTS
OF CLIMATE CHANGE

Climate change reinforces many existing causes of morbidity and mortality and may add new health risks in regions that historically have been relatively free of them (see table 16). Complicating adaptation policy will be the fact that net health impacts of climate change are not straightforward. The incremental health risk of something a pervasive as climate change is very difficult to measure, and sometimes even works in conflicting ways. As climate change increases heat-related health risks, for instance, it will reduce cold-related health risks. Air quality impacts of climate change also depend upon complex changes in the atmosphere, such as reaction rates, relative humidity, wind speed, and cloud cover.

Some of the impacts listed in table 16 are still not well understood. These include, for example, how changes in atmospheric chemistry and wildfires affect small-diameter particulate matter (PM 2.5) exposures and the relationship among climate change, natural disasters, and health risks. Others, such as changes in ozone levels and a rise in the frequency, intensity, and duration of heat waves, are better understood. Even in cases where the health risks of climate change are better characterized, however, the potential economic impacts of climate change on health remain very uncertain.

Ozone

Ozone, one of six "criteria" pollutants regulated by the Environmental Protection Agency (EPA), forms in the atmosphere from two precursors: nitrogen oxides (NO_x) and volatile organic

TABLE 16

Potential Climate Change Impacts on California's Public Health

Climate impact	Health impact
Increase in temperatures	Increase in ozone levels
	Increase in small-diameter particulate matter (PM 2.5) levels
	Change in the geography of vector-borne diseases
	Change in the geography of allergens and allergic diseases
	Decrease in seasonal extreme cold temperatures
Changes in atmospheric water vapor	Change in ozone levels
	Change in particulate density
	Change in aerosol pollution retention
Change in precipitation	Change in the geography of water- and vector-borne diseases
	Increase in flooding
Increase in temperature extremes	Increase in heat waves
Increase in wildfires	Increase in ozone and PM 2.5 levels

chemicals (VOCs).[5] Ground-level ozone is a public health concern, exacerbating respiratory illnesses and, in a limited number of cases, leading to premature death.[6] Active children are typically the most at-risk group with respect to higher ozone levels because they are more frequently outdoors during the high ozone season, which typically extends from May to October.[7] Ozone is a particularly difficult problem for California because of the state's distinctive meteorology. The only two areas of the United States where EPA has designated "extreme" levels of one-hour and eight-hour ozone concentrations are both in California.[8]

Changes in climate affect ozone concentrations in a number of ways. Rising temperatures speed up reaction rates for both ozone formation and removal. Higher temperatures also increase NO_x emissions from power plants and vehicles and biogenic VOC emissions, which, in tandem, could lead to higher ozone concentrations.[9] Changes in atmospheric water vapor and air circulation patterns may increase or decrease local ozone levels.[10] The net effect of these changes is not straightforward, but the bulk of existing research, both for California and globally, suggest that climate change will tend to foster higher ozone concentrations (see table 17).

Regulatory efforts to limit NO_x and VOC emissions will likely continue to achieve relative reductions in ozone levels, but climate change will reduce the effectiveness, and cost-effectiveness, of these efforts. Additionally, background ozone levels, formed from NO_x and VOC emissions outside California, are increasing and may make it difficult to achieve state ozone standards in some regions.[11] Climate change and a sustained increase in background ozone will shift the control cost curve for ozone leftward, raising the cost to meet any given ozone standard. If the control cost curve for ozone is steeply sloped, as in figure 9, returning to current levels of ozone control may lead to such a sharp rise in costs as to render them cost-ineffective.

In addition to increased costs of control, climate change–induced escalation of ozone levels would escalate health costs. Based on the estimates in table 17, a 10 to 20 percent decrease in ozone level reductions would lead to increased societal costs, both in terms of health care costs and the value of lost or attenu-

TABLE 17

Projected Increase in High Ozone Days Resulting from Climate Change

Region	Climate model–emissions scenario	Time frame	Metric	Impact	Source
San Joaquin Valley Air Basin, South Coast Air Basin	GFDL–A2 GFDL–B1	2050	Days per year over 90 parts per billion[a]	22–30 days per year 6–13 days per year	Mahmud et al.
San Joaquin Valley (Visalia), South Coast Air Basin (Riverside)	GFDL–A2 GFDL–B1	2100	Percent increase in days per year over 90 parts per billion	75%–85% 25%–35%	Drechsler et al.
Sacramento, Fresno, area south and east of San Francisco Bay	Range of temperatures and meteorological conditions	2050	Percent change in weekday maximum ozone concentration	Climate only: +3%–10% Climate + emission reductions: (−9%)–(+1%)	Steiner et al.

NOTE: GFDL is a low-sensitivity climate model. Climate sensitivity is the responsiveness of the earth's near-surface air temperature to changes in radiative forcing. B1 is a low greenhouse gas emissions scenario; A2 is a business-as-usual emissions scenario.

SOURCES: Abdullah Mahmud et al., "Statistical Downscaling of Climate Change Impacts on Ozone Concentrations in California," *Journal of Geophysical Research* 113 (208): 1–12; Drechsler et al., "Public Health–Related Impacts of Climate Change in California," California Climate Change Center White Paper, CEC-500-2005-197-SF, 2006; Allison L. Steiner et al., "Influence of Future Climate and Emissions on Regional Air Quality in California," *Journal of Geophysical Research* 111 (2006): 1–22.

[a] For reference, in 2010 the Southern California Air Basin and San Joaquin Valley Air Basin had 85 and 59 days, respectively, above California's 90-parts-per-billion 1-hour ozone standard. Data are from the California Air Resources Board's Air Quality Data Statistics database, www.arb.ca.gov/adam.

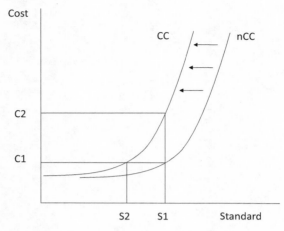

Figure 9. Illustrative example of the impact of climate change on ozone cost control. Climate change (CC) shifts the "no climate change" ozone control cost curve (nCC) left to CC, decreasing the standard met by current efforts from S_1 to S_2 at cost C_1. In this example, maintaining the S_1 standard would roughly double control costs. Here, S_1 and S_2 would be the inverse of concentration-based standards.

ated lives, on the order of hundreds of millions to billions of dollars per year.[12]

Heat Waves

Heat waves, though lacking a universal definition, generally mean sustained intervals of hot temperatures. Extreme heat is adverse to public health, increasing the risks of life-threatening conditions like heat stroke, heart attack, and severe dehydration, particularly among the elderly, chronically ill, children, some ethnic minority groups, and farm workers. A study across nine

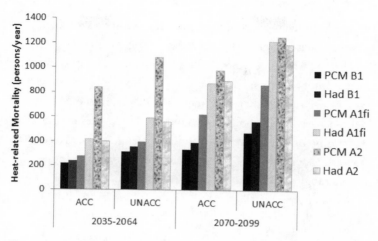

Figure 10. Projected heat-related mortality in Fresno, Los Angeles, Sacramento, San Bernardino, and San Francisco, 2035–2064 and 2070–2099, for acclimatization (ACC) and no acclimatization (UNACC). Source: Dreschler et al., "Public Health–Related Impacts of Climate Change in California," California Climate Change Center White Paper, CEC-500-2005-197-SF, 2006.

counties in California concluded that a 10°F (6°C) increase in average temperatures corresponded to a 2.3 percent increase in heat-related mortality.[13]

In California, the scope and severity of heat-related mortality risk is already nontrivial. Coroners reported 142 deaths resulting from the state's 2006 heat wave.[14] Moreover, because of the difficulty of attributing cause of death to heat, these estimates may underreport actual deaths by a factor of as much as three.[15] Drechsler et al. estimate that actual heat-related mortality in Fresno, Los Angeles, Sacramento, San Bernardino, and San Francisco averaged 261 persons per year during the 1990s.[16]

Climate change is reliably expected to increase the frequency, duration, and intensity of heat waves.[17] Drechsler et al. project that climate change would increase the frequency of days with extreme heat twofold to fourfold, depending on climate scenario, by the end of the century.[18] They estimate that without public health interventions, this increased likelihood of extreme heat could raise the number of heat-related deaths in the five metropolitan areas listed above by around 70 to 1,100 persons per year, depending on acclimatization and adaptation (fig. 10). Across the state, the social costs of heat-related mortality at the higher end of this range would be on the order of billions to tens of billions of dollars per year.[19]

ADAPTATION OPTIONS

California already has a strong public health system that includes a number of institutions essential for responding to a changing climate, such as heat emergency planning, disease monitoring and control programs, and sophisticated air quality monitoring and research. Adaptation in this area will require gradually strengthening these institutions, particularly monitoring capabilities and response capacity for a higher frequency and variability of weather-related impacts on public health.

The spectrum of public health responses to climate change extends from prevention, or reducing the source of disease (heat and pollution), to treatment, or dealing with the consequences. For vector and heat-related disease, prevention is likely to be highly cost-effective. Drechsler et al., for instance, estimate that the cost of regional emergency heat warning systems, intervention plans, and intervention activities would be around $1 million.[20] For air pollution, and ozone in particular, the costs of prevention range from less expensive monitoring and information systems that are designed to reduce exposure, to more expensive pollution control equipment. The challenge for public agencies will be to find the right mix of adaptation measures to reduce health risks at an acceptable cost.

As with the other risks we have discussed, equity implications should figure into this calculus. The public health impacts of climate change, and in particular the two we discussed in greater depth here, affect different segments of the population. Those most at risk from heat waves in California, for instance, are infants, African Americans, and people over the age of sixty-five.[21] Exposure to ozone in Southern California appears to be greatest among higher-income Caucasians living in lower-density areas,[22] although exposure must be combined with health status to fully assess health impacts. Effective targeting of policy, matching health response strategies to affected populations, can significantly improve the effectiveness of required health interventions and reduce their public and private costs.

Public health has a more direct link to climate change mitigation than many other sectors, although this link is not always straightforward. Reducing the use of natural gas in power generation and the use of petroleum products in transportation could reduce NO_x emissions and ozone concentrations, but some low-carbon technologies, such as biodiesel, could increase NO_x emissions without appropriate control technologies.[23] Some adaptation behavior may contradict mitigation goals. For example, air-conditioning to lessen the effect of heat waves will increase peak electricity demand, which will, all else being equal, increase the use of inefficient and high-polluting natural gas plants. The California Air Resources Board, which implements AB 32 and manages the state's air quality, is well positioned to ensure that mitigation and air quality goals are aligned, but it is less able to ensure that broader public health adaptation measures are aligned with the state's mitigation goals and vice versa. This diffusion of interrelated responsibilities across state agencies is but one example of the need for determined interagency coordination to develop consistent plans for climate mitigation and adaptation.

Looking at the wider context, climate change has already begun to exacerbate public health risks around the world, and in California air quality and temperature will present the most serious challenges. California already has the worst air quality of all fifty United States. Official estimates of the public cost of current air pollution are high, ranging from $36 billion to $136 billion annually. Most of this is actuarial valuation of premature death, but hundreds of millions of dollars are also spent on air pollution–related hospital visits. Lost work and school days and reduced activity all increase the cost of pollution to society. Without continued, and potentially more expensive, pollution control efforts, climate change will increase all these social costs.

Effective government intervention could significantly reduce health-related impacts from climate change. Controlling criteria pollutant emissions is the most potent option for reducing the pollution-related impacts of climate change. Relatively simple strategies could significantly reduce mortality during heat waves in California. These include early warning systems, public education, cooling centers, and air-conditioning. Longer-term solutions for air quality are also likely to give rise to a host of other considerations, from transportation and land use planning to fuel choices to greater attention to environmental justice issues. Among older and poorer groups, adaptive capacity is also significantly lower than in the population at large. Given the potentially high rates of risk and relatively low costs of intervention, there is an obvious role for public agency in reducing heat-related mortalities.

CHAPTER NINE

Revelation or Revolution?

How California Can Adapt and Even Benefit from Climate Change

This book examines the economics of climate change in California based on a comprehensive review of relevant research across seven sectors of the state economy. Like the seven seals on the Book of Revelation, each of these cases reveals a different dimension of one overarching threat to our well-being, changes in the natural environment that will have profound and lasting adverse consequences if we do not respond appropriately. Climate change is not only inevitable, it is already happening. Unlike a prophecy, however, its consequences are not inevitable and they can be significantly mitigated by human and institutional adaptation. We can choose between viewing climate change as a curse or as an opportunity, depending on our willingness to adapt.

Our comprehensive review of existing research shows that if we do nothing, changes in the natural world will put our health, assets, and livelihoods at serious risk. The most expensive climate impacts for California are likely to be to human health and property. There are varying degrees of public and private risk across sectors, and the main impact on the general public will be through changes in prices (e.g., food, water, and energy may become more expensive relative to widgets). Overall, annual costs to the state will probably be relatively small, measured as a share of gross state product, as California produces more widgets than food, water, and energy. However, beneath the smooth veneer of economic aggregates and averages there will be significant trade-offs and distributional impacts—winners and losers. Discovering who and where these people are will probably arouse the public interest, but it would be much less expensive to anticipate these adjustments (e.g., mortality from heatstroke) than to react to them. As we emphasize, huge wealth transfers—from the public to the private, from producer to consumer, from producer to producer, and from consumer to consumer—could potentially result from climate change. These will seriously complicate the politics of adaptation (e.g., if the state decides to continue to be an insurer of last resort for fires and subsidize insurance for Malibu homes). Moreover, the economic impacts could be highly uneven across the state, leading to stakeholder competition for adaptation resources.

The good news is that the costs of climate adaptation are highly dependent on policy choices, and, as we argue above, there are many choices open to California. The bad news is that the most cost-effective and equitable solutions require strong institutions (e.g., for water trading, conjunctive management of surface and groundwater, public health information systems, risk-based insurance pricing, and demand-side management) that only partially resemble their counterparts today. Moreover, expensive solutions are often capital intensive (e.g., new surface reservoir capacity, desalinization, publicly backed insurance, new gas power plants), are in some ways extensions of the status quo, and have interest groups attached to them.

Because of this combination of ill-prepared institutions and their legacy relationships, the truth is that California is woefully unready to adapt to climate change. Worse, the state's lack of readiness is most acute in ways that are not captured in the current climate adaptation discourse, focused as it is on science and engineering problems. Even on technical grounds, the state needs to reappraise its capacity. Over the last century California's infrastructure, institutions, and politics have evolved in a small window of low climate variability; none are up to the task of handling more significant natural climate variability, let alone increasing variability. Average cost/risk measures are also misleading because the downside costs are much greater than the upside potential of investments to avert damages. As with most disaster preparedness, we need to anticipate low-probability events with very high cost, and this means overcommitting public resources most of the time, as fire departments do. Public decision makers need maturity to make and sustain such commitments against the cycles of economic growth and political seasonality.

The real reason to start thinking about adaptation now is not so that we can actually begin to build more infrastructure such as seawalls and reservoirs, but rather so that we can figure out what the institutional and political barriers are to dealing with not just rising average temperatures but also higher natural and anthropogenic climate variability, and we can figure out ways to remove or reduce those barriers. Essential decisions, including those about the nature of public-private participation, risk sharing, and fiscal mechanisms, take time in a democracy, and they can often be politically and demographically intergenerational. Determination to reach these decisions is essential, however, because they will determine whether adaptation is cheap or very expensive. To support this process, California's adaptation research agenda needs to be refocused and intensified. In this way it can address more diverse, short-term processes and impacts and identify barriers to, and solutions for, cost-effective and timely adaptation.

Beyond our own adaptation needs, there also lies a momentous opportunity. Climate change presents a global challenge that will require innovation on a scale comparable to any period since (and perhaps including) the industrial revolution. Space programs were national and narrowly focused, the Green Revolution addressed only the agrofood supply chain, and even the two world wars concentrated most of their innovation potential in a few activities (e.g., ordnance, logistics, and transport). To address all the resource, livelihood, and habitat implications of climate change will require much more than money and willing, robust public institutions. Because of the technical nature of climate risk's origins and impacts, effective adaptation will require a knowledge-intensive revolution in our understanding of how to live in the natural world and use its resources more sustainably and cooperatively. What better place for this than California, the birthplace of the last two great knowledge-intensive industries, information technology and biotech? Using our own experience, facilitated by state policy leadership, to incubate the first generation of adaptation strategies and technologies, California can capture global competitive opportunities that last for generations, doubly securing our prosperity and maintaining the luster of the Golden State.

NOTES

INTRODUCTION

1. California's policy initiative half a decade ago was also supported by high-quality research on the consequences of inaction, summarized in Katharine Hayhoe, Daniel Cayan, Christopher B. Field, Peter C. Frumhoff, Edwin P. Maurer, Norman L. Miller, Susanne C. Moser, Stephen H. Schneider, Kimberly Nicholas Cahill, Elsa E. Cleland, Larry Dale, Ray Drapek, R. Michael Hanemann, Laurence S. Kalkstein, James Lenihan, Claire K. Lunch, Ronald P. Neilson, Scott C. Sheridan, and Julia H. Verville, "Emissions Pathways, Climate Change, and Impacts on California," *Proceedings of the National Academy of Sciences* 101 (2004): 12422–27.

CHAPTER ONE

1. José Goldemberg et al., "Introduction: Scope of the Assessment," in *Climate Change 1995: Economic and Social Dimensions of Climate Change,* ed. J.P. Bruce et al., Contribution of Working Group III to the Second Assessment Report of the Intergovernmental Panel on Climate Change (Cambridge: Cambridge University Press, 1996).

2. UNEP and McKinsey, "Shaping Climate Resilient Develop-

ment: A Framework for Decision Making," Economics of Climate Adaptation Working Group, 2009.

3. Michael Hanemann, "Observations on the Economics of Adaptation: Uncertainty and Timing," presentation to the OECD Workshop on the Economics of Adaptation, April 7–8, 2008, OECD, Paris; Shardul Agrawala and Samuel Fankhauser, *Economic Aspects of Adaptation to Climate Change: Costs, Benefits and Policy Instruments* (Paris: OECD, 2008).

4. World Bank, "Overview: Changing the Climate for Development," in *World Development Report 2010: Development and Climate Change* (Washington, DC: World Bank, 2009).

5. James E. Neumann et al., "Market Impacts of Sea Level Rise on California Coasts," Appendix XIII in Tom Wilson et al., "Global Climate Change and California: Potential Implications for Ecosystems, Health, and the Economy," PIER Report 500-03-058CF, 2003.

6. The underlying theory is summarized in Kenneth J. Arrow et al., "Intertemporal Equity, Discounting, and Economic Efficiency," in *Climate Change 1995: Economic and Social Dimensions of Climate Change,* ed. J. Bruce, H. Lee, and E. Haites (Cambridge: Cambridge University Press, 1996). A detailed discussion of the discounting issue can be found in Yale Symposium, "Yale Symposium on the Stern Review," Yale Center for the Study of Globalization, Yale University, February 2007.

7. According to Ramsey, the standard discount rate, $r = \delta + \eta g$, comprises a "pure" rate of time preference (δ), a coefficient of relative risk aversion (η), and the per capita growth rate of consumption (g). A low value of the rate of pure time preference represents preference for intergenerational equity, while a high value of the coefficient of relative risk aversion implies equity over space and time.

8. Everett Crosby, "Fire Prevention," *Annals of the American Academy of Political and Social Science* 26 (1905): 224–38. More generally, Paul Krugman described moral hazard as "any situation in which one person makes the decision about how much risk to take, while someone else bears the cost if things go badly." See Paul Krugman, *The Return of Depression Economics and the Crisis of 2008* (New York: W. W. Norton, 2009).

9. Even when the cost is known, as it is for interest payments on

deficit spending, the possibility of deferring it to future generations creates moral hazard.

CHAPTER TWO

1. From 2000 to 2010, agriculture, forestry, fisheries, and hunting accounted for an average of 1.36 percent of California's real GSP. Data are from the Bureau of Economic Analysis (BEA) website, "Regional Data," http://bea.gov/iTable/iTable.cfm?reqid=70&step=1&isuri=1&acrdn=3.

2. UC Agricultural Issues Center (UCAIC), "The Measure of California Agriculture: Highlights," 2009.

3. California Department of Food and Agriculture, "Agricultural Statistical Review," California Agricultural Resource Directory 2010–2011, 2011.

4. American Forest & Paper Association, "Forest & Paper Industry at a Glance: California," 2011.

5. See David B. Lobell, Wolfram Schlenker, and Justin Costa-Roberts, "Climate Trends and Global Crop Production Since 1980," *Science* (2011): 1–5; Wolfram Schlenker and David B Lobell, "Robust Negative Impacts of Climate Change on African Agriculture," *Environmental Research Letters* 5 (2010): 1–8.

6. The California Air Resources Board (ARB) estimates that, producing 28.1 million tons of CO_2 equivalent (tCO_2e) in 2008, agriculture accounted for 6 percent of the state's greenhouse gas emissions. See California Air Resources Board, "Trends in California Greenhouse Gas Emissions for 2000 to 2008—By Category as Defined in the Scoping Plan," www.arb.ca.gov/cc/inventory/data/tables/ghg_inventory_trends_00-08_2010-05-12.pdf.

7. See David Roland-Holst and David Zilberman, "How Vulnerable Is California Agriculture to Higher Energy Prices?" *Agricultural and Resource Economics Update,* University of California Giannini Foundation, 2006; Daniel A. Sumner and John Thomas Rosen-Molina, "Impacts of AB32 on Agriculture," in *Agricultural and Resource Economics Update,* University of California Giannini Foundation, 2010.

8. For a discussion of key differences among statistical approaches, see W. Schlenker, W. Michael Hanemann, and Anthony C. Fisher, "Will U.S. Agriculture Really Benefit from Global Warming? Accounting for Irrigation in the Hedonic Approach," *American Economic Review* 95 (2005): 395–406.

9. J. Hatfield et al., "Agriculture," in *The Effects of Climate Change on Agriculture, Land Resources, Water Resources, and Biodiversity in the United States*, ed. Backlund et al. A Report by the U.S. Climate Change Science Program and the Subcommittee on Global Change Research. Washington, D.C.

10. David B. Lobell, Angela Torney, and Christopher B. Field, "Climate Extremes in California Agriculture," California Climate Change Center Draft Paper, CEC-500-2009-040-D, 2009.

11. The National Oceanic and Atmospheric Administration (NOAA) estimates that crop damages in California averaged $322 million a year between 1996 and 2010 (in 2005 dollars), with a high of $2 billion in 1998. Data are from NOAA, Office of Climate, Water, and Weather Services, "Natural Hazard Statistics," www.nws.noaa.gov/om/hazstats.shtml. We assume that the NWS data are in nominal dollar terms and have deflated them to 2005 dollars using the IMF's deflator for the United States. These estimates are different from those made by the NOAA's National Climatic Data Center (NCDC), which estimates the cost of California's 1998 freeze at $3.2 billion. By the NCDC's estimate, the largest disaster in California's recent history was a $5.5 billion freeze in 1990. Data are from NCDC, "1980–2010 Billion Dollar U.S. Weather Disasters," www.ncdc.noaa.gov/img/reports/billion/disasters2010.pdf. Percentages of agricultural revenues are based on the National Agricultural Statistics Service's (NASS's) Census of Agriculture, where total commodity sales of California agriculture were $23,280 million in 1997, $25,737 million in 2002, and $33,885 million in 2007 (in constant dollars).

12. Lobell et al., "Climate Extremes in California Agriculture."

13. See John T. Trumble and Casey D. Butler, "Climate Change Will Exacerbate California's Insect Pest Problems," *California Agriculture* 63 (2009): 73–78; Dennis Baldocchi and Simon Wong, "An Assess-

ment of the Impacts of Future CO_2 and Climate on Californian Agriculture," California Climate Change Center White Paper, CEC-500-2005-187-SF, 2006.

14. See Richard Howitt and Randall G. Mutters, "Economic Assessment of Acid Deposition and Ozone Damage on the San Joaquin Agriculture," California Air Resources Board Research Note 94-18, 1994; Mark A. Delucchi, James Murphy, Jin Kim, and Donald R. McCubbin, "The Cost of Crop Damage Caused by Ozone Air Pollution from Motor Vehicles," Institute of Transportation Studies Report UCD-ITS-RR-96-3 (12), 1996.

15. In addition to the studies listed in table 2, see David Lobell and Christopher Field, "California Perennial Crops in a Changing Climate," California Climate Change Center Final Paper, CEC-500-2009-039-F, 2009.

16. California Department of Food and Agriculture, "Agricultural Statistical Review."

17. John J. Battles et al., "Climate Change Impacts on Forest Growth and Tree Mortality: A Data-driven Modeling Study in the Mixed Conifer Forest of the Sierra Nevada, California," *Climatic Change* 87 (2008): S193–S213; John Battles et al., "Projecting Climate Change Impacts on Forest Growth and Yield for California's Sierran Mixed Conifer Forests," California Climate Change Center Final Paper, CEC-500-2009-047-F, 2009.

18. Henk Visscher, Mark A. Sephton, and Cindy V. Looy, "Fungal Virulence at the Time of the End-Permian Biosphere Crisis?" *Geology* 39 (2011): 883–86.

19. Robert Mendelsohn, "A California Model of Climate Change Impacts on Timber Markets," Appendix XII in *Global Climate Change and California: Potential Implications for Ecosystems, Health, and the Economy,* ed. T. Wilson, L. Williams, J. Smith, and R. Mendelsohn, PIER Publication No. 500-03-058CF.

20. Lee Hannah et al., "The Impact of Climate Change on California Timberlands," California Climate Change Center Final Paper, CEC-500-2009-045-F, 2009.

21. California Department of Finance, "California County Profiles," www.dof.ca.gov/HTML/FS_DATA/profiles/pf_home.php.

22. Gretta T. Pecl and George D. Jackson, "The Potential Impacts of Climate Change on Inshore Squid: Biology, Ecology and Fisheries," *Reviews in Fish Biology and Fisheries* 18 (2008): 373–85.

23. Ibid.

24. United States Global Change Research Program, "Impacts and Adaptation," in *Climate Action Report 2002* (Washington, DC: USGCRP, 2002).

25. Edward A. Parson et al., "Potential Consequences of Climate Variability and Change for the Pacific Northwest," in *Climate Change Impacts on the United States*, ed. Jerry M. Melillo et al. (Cambridge: Cambridge University Press, 2009).

26. The PDO was described only recently and is still poorly understood. See Parson et al., "Potential Consequences of Climate Variability."

27. Merrill et al. report that the number of on-farm extension advisors and specialists has fallen 40 percent since the early 1990s. See Jeanne Merrill, Renata Brillinger, and Allyse Heartwell, "Ready ... Or Not? An Assessment of California Agriculture's Readiness for Climate Change," California Climate and Agricultural Network Report, March 2011.

28. California agricultural labor data are from the BEA website, "Regional Data." The nine counties are Colusa, Imperial, Kings, Madera, Monterey, Riverside, San Benito, Tulare, and Yuba. Data are from California Department of Finance, "California County Profiles."

29. John J. Battles et al., "Climate Change Impact on Forest Resources," California Climate Change Center White Paper, CEC-500-2005-193-SF, 2006.

CHAPTER THREE

1. For more details on these projections, see, for example, Nathan T. Van Rheenan et al., "Potential Implications of PCM Climate

Change Scenarios for Sacramento–San Joaquin River Basin Hydrology and Water Resources," *Climatic Change* 62: 257–81.

2. Ellen Hanak et al., *Managing California's Water: From Conflict to Reconciliation* (San Francisco: Public Policy Institute of California, 2011), 160.

3. This average 50 percent share of environmental water use is misleading, as environmental use ranges from much higher in the wetter, northern parts of the state to much lower in the drier, central and southern parts of the state. See ibid., 88.

4. All of the shares in this paragraph are from ibid., 87.

5. Daniel R. Cayan, Edwin P. Maurer, Michael D. Dettinger, Mary Tyree, and Katharine Hayhoe, "Climate Change Scenarios for the California Region," *Climatic Change* 87 (2008): S21–S42.

6. All of the results in table 6 are based on two water sector models: the California Value Integrated Network (CALVIN) model and the California Water Resources Simulation (CalSim) model. CALVIN is an engineering-economic optimization model, whereas CalSim is a water resources simulation model. The Hanemann et al. results are based on an ex post analysis of CalSim outputs.

7. This estimate assumes that California has 20.7 million households in 2050, based on a population projection of 59,507,876 people and an average household size of 2.87. Population data are from State of California, Department of Finance, "Population Projections for California and Its Counties 2000–2050, by Age, Gender and Race/Ethnicity," July 2007. Household size data are from the U.S. Census Bureau, "Fact Sheet, California, Census 2000 Demographic Profile Highlights."

8. Based on Hanak et al., *Managing California's Water,* 96.

9. California Department of Water Resources, "California Water Plan Update," Bulletin 160-09, 2009.

10. State of California, Department of Finance, "Population Projections."

11. Ellen Hanak and Jay Lund, "Adapting California's Water Management to Climate Change," in *Preparing California for a Changing Climate,* ed. Louise Bedsworth and Ellen Hanak (San Francisco: Public Policy Institute of California, 2008).

12. Stacy K. Tanaka et al., "Climate Warming and Water Management Adaptation for California," *Climatic Change* 76 (2006): 361–87.

13. See Scott Stine, "Extreme and Persistent Drought in California and Patagonia during Mediaeval Time," *Nature* 369 (1994): 546–49. Stine argued that two major droughts continued uninterrupted for more than a century, but Graham and Hughes provide evidence that these long drought periods were punctuated by years with higher rainfall. See Nicholas E. Graham and Malcolm K. Hughes, "Reconstructing the Mediaeval Low Stands of Mono Lake, Sierra Nevada, California, USA," *Holocene* 17 (2007): 1197–1210.

14. Larry Benson, Michaele Kashgarian, Robert Rye, Steve Lund, Fred Paillet, Joseph Smoot, Cynthia Kester, Scott Mensing, Dave Meko, and Susan Lindström, "Holocene Multidecadal and Multicentennial Droughts Affecting Northern California and Nevada," *Quaternary Science Reviews* 21 (2002): 659–82.

15. Hanak et al., *Managing California's Water,* 149.

16. Collateral damage to agriculture and property is considered in other chapters.

17. Hanak and Lund, "Adapting California's Water Management," 10.

18. Irrigated pasture and field crops accounted for more than 60 percent of California's agricultural water use but less than 15 percent of gross agricultural revenues in 2005. Field crops in particular remain viable because of crop subsidies, water subsidies, and seniority of water rights. See Hanak et al., *Managing California's Water,* 92–93.

19. See, for instance, Richard Howitt, "Delta Dilemmas: Reconciling Water-Supply Reliability and Environmental Goals," *Agricultural and Resource Economics Update* 10 (2007): 1–4.

20. Such was the conclusion of the Delta Vision Foundation's 2011 Report Card. The Delta Vision Foundation is a monitoring and evaluation body created by the Delta Vision Blue Ribbon Task Force, an independent advisory organization set up by the state governor. See Delta Vision Foundation, "2011 Delta Vision Report Card," June 2011 Report.

21. P. C. D. Milly et al., "Stationarity Is Dead: Whither Water Management?" *Science* 319 (2008): 573–74.

22. This more decentralized model of governance contrasts with California's electricity sector, which is dominated by large investor-owned utilities and where regional integrated planning has a long history. Which model is more effective, efficient, and equitable for adapting to climate change is an open question.

23. Hanak et al., *Managing California's Water*, 274.

24. Hanak and Lund, "Adapting California's Water Management," 28–29.

25. A number of individually initiated trades worth several hundred million dollars have been offered in recent years, with no takers in the agricultural community.

CHAPTER FOUR

1. For a general overview, see Pat Perez, "Potential Impacts of Climate Change on California's Energy Infrastructure and Identification of Adaptation Measures," California Energy Commission Staff Paper, CEC-150-2009-00, 2009. For an overview of climate adaptation in the electricity sector, see Edward Vine, "Adapting California's Electricity Sector to Climate Change," in *Preparing California for a Changing Climate*, ed. Louise Bedsworth and Ellen Hanak (Public Policy Institute of California, 2008).

2. The two major studies on meeting longer-term greenhouse gas targets in California have concluded that, with current technologies, significant electrification will be necessary to meet an 80 percent 2050 target. See Energy and Environmental Economics, "Meeting California's Long-term Greenhouse Gas Reduction Goals," Consultant Report for Hydrogen Energy International LLC, 2009; Lawrence Berkeley National Lab, "California's Energy Future: Targeting 80% Emissions Reduction," http://carboncycle2.lbl.gov/event_20110420.html.

3. Thomas P. Hughes, *Networks of Power: Electrification in Western*

Society, 1880–1930 (Baltimore, MD: Johns Hopkins University Press, 1983), 262–84.

4. Aspen Environmental Group and M. Cubed, "Potential Changes in Hydropower Production from Global Climate Change in California and the Western United States," California Energy Commission Consultant Report, CEC-700-2005-010, 2005.

5. Gross system power is defined as total consumption plus imports. Electrons are not differentiated on the grid, and thus the composition of imports is unknown and must be estimated. Data here are from the California Energy Commission's (CEC's) online Energy Almanac, "Total Electricity System Power," http://energyalmanac.ca.gov/electricity/total_system_power.html.

6. Vine, "Adapting California's Electricity Sector," 6.

7. Aspen Environmental Group and M. Cubed, "Potential Changes in Hydropower Production from Global Climate Change in California and the Western United States," 3–4, 31.

8. At an overall thermal efficiency of 40 percent and an IPCC default emission factor of 56.1 kg of CO_2 per giga-joule (GJ), natural gas–fired generation has an emission factor of 0.5 kg of CO_2 per kilowatt-hour (kWh), whereas coal-fired generation with an efficiency of 33 percent and an IPCC default emission factor of 96.1 kg of CO_2 per GJ has an emission factor of 1.0 kg of CO_2 per kWh. Howarth et al. argue that accounting for fugitive methane emissions significantly raises the emission factor of natural gas combustion, particularly shale gas. See Robert W. Howarth, Renee Santoro, and Anthony Ingraffea, "Methane and the Greenhouse-Gas Footprint of Natural Gas from Shale Formations," *Climatic Change* 106 (2011): 679–90.

9. Philip W. Mote and Eric P. Salathe Jr., "Future Climate in the Pacific Northwest," *Climatic Change* 102 (2010): 29–50.

10. Tanaka et al. demonstrate this point for low-elevation dams. See Stacy K. Tanaka et al., "Climate Warming and Water Management Adaptation for California," *Climatic Change* 76 (2006). Madani and Lund and Vicuna et al. demonstrate it for high-elevation dams. See Kaveh Madani and Jay R. Lund, "Estimated Impacts of Climate Warming on California's High-Elevation Hydropower," *Climatic Change* 102 (2009):

521–38; S. Vicuna et al., "Climate Change Impacts on High Elevation Hydropower Generation in California's Sierra Nevada: A Case Study in the Upper American River," *Climatic Change* 87 (2008): S123–S137.

11. Matthew S. Markoff and Alison C. Cullen, "Impact of Climate Change on Pacific Northwest Hydropower," *Climatic Change* 87 (2008): 451–69.

12. From Madani and Lund, this estimate assumes that high-elevation dams make up around 75 percent of in-state generation and that 20 percent of that generation is lost by the end of the century; from Chung et al., that around 15 percent of in-state generation from low-elevation dams is lost by the end of the century; based on Markoff and Cullen, that hydropower exports from the Pacific Northwest to California are reduced by 50 percent by the end of the century; and, based on a lack of estimates, that Colorado River hydropower exports to California are unchanged. See Madani and Lund, "Estimated Impacts of Climate Warming"; Francis Chung et al., "Using Future Climate Projections to Support Water Resources Decision Making in California," California Climate Change Center Report, CEC-500-2009-052-F, 2009; Markoff and Cullen, "Impact of Climate Change." From CEC, "Total Electricity System Power," in-state, Pacific Northwest, and Southwest hydropower accounted for 72, 23, and 5 percent, respectively, of total hydropower in California from 2002 to 2009. Together, these imply that hydropower availability would be (75% × 80% + 25% × 85%) × 72% + 50% × 23% + 5% = 75% of historical levels.

13. For replacement quantity, this calculation uses average 2002–2009 hydropower levels of 48 terawatt-hours, from CEC, "Total Electricity System Power." For replacement price, it uses an average cost of $0.14 per kWh, based on the long-term market price referent (MPR) set in 2009 as a proxy for longer-term costs. The MPR is a proxy used by the California Public Utilities Commission to reflect the long-run cost of a 500-megawatt combined-cycle gas turbine, including both long-term costs and time-of-delivery values. See "Public Utilities Commission of the State of California Energy Division Resolution E-4298," http://docs.cpuc.ca.gov/PUBLISHED/FINAL_RESOLU-TION/111386.htm.

14. Because the California electricity sector still maintains a regulated utility model on the retail side, expenditures are a useful proxy for costs. California residents, businesses, and governments spent roughly $34 billion on electricity in 2009, of which $1.7 billion would be 5 percent. By the end of the century, as revenues and costs rise, $1.7 billion would be a much smaller share of electricity system costs. For expenditure data, see Energy Information Administration, "Table F22: Electricity Price and Expenditure Estimates, 2009," www.eia.gov/emeu/states/hf.jsp?incfile=sep_fuel/html/fuel_pr_es.html.

15. This calculation uses data from the California Power Plant Database, http://energyalmanac.ca.gov/electricity/index.html.

16. Vine, "Adapting California's Electricity Sector," 3.

17. B. Lebassi et al., "Impacts of Climate Change in Degree Days and Energy Demand in Coastal California," *Journal of Solar Energy Engineering* 132 (2010): 031005-7.

18. Ibid., 031005-8.

19. More specifically, 26 percent of the U.S. Census Bureau's projected growth in population by 20.4 million persons between 2010 and 2050 is in the eight San Joaquin Valley counties of Fresno, Kern, Kings, Madera, Merced, San Joaquin, Stanislaus, and Tulare; 20 percent is in the three inland Southern California counties of Imperial, Riverside, and San Bernardino. These eleven counties accounted for an estimated 22 percent of the state's population in 2010. Data are from Rand California, "California Population Projections by Ethnicity and Age," http://ca.rand.org/stats/popdemo/popproj.html.

20. Only in Kings County (2,112 kWh per person) and San Bernardino County (2,108 kWh per person) was per capita electricity consumption below the state average (2,258 kWh per person) in 2010. Electricity data are from the California Energy Commission, Energy Consumption Data Management System, www.ecdms.energy.ca.gov/elecbycounty.aspx.

21. The Auffhammer and Aroonruengsawat projections are for residential use only. This estimate assumes that commercial electricity use has the same climate elasticity as residential use. Residential and commercial electricity use in California averaged 146.1 billion kWh

per year from 1980 to 2000. Costs are calculated using the MPR from note 13, above. Cost increases are likely to be higher than this estimate because rising temperatures are likely to increase electricity use when electricity is most expensive, whereas the MPR is based on an average cost. Residential and commercial electricity use data are from EIA, State Energy Data System, http://205.254.135.24/state/seds/.

22. This is much smaller, for example, than the growth of gasoline prices over the last fifty years, during which California per capita vehicle miles traveled increased substantially.

23. For instance, California's Global Warming Solutions Act (AB 32) established the Economic and Technology Advancement Advisory Committee (ETAAC) to provide guidance on technologies for reducing greenhouse gas emissions. ETAAC's December 2009 final report does not mention the word *adaptation*. See Economic and Technology Advancement Advisory Committee, "Advanced Technology to Meet California's Climate Goals: Opportunities, Barriers & Policy Solutions," www.arb.ca.gov/cc/etaac/meetings/ETAACAdvancedTechnologyFinalReport12-14-09.pdf.

CHAPTER FIVE

1. The air, rail, water, truck, and transit and ground passenger transportation sectors generated $21 billion of value added in 2009. Data are from the Bureau of Economic Analysis (BEA) website, "Regional Data," http://bea.gov/iTable/iTable.cfm?reqid=70&step=1&isuri=1&acrdn=3.

2. In 2008, the ports of Los Angeles, Long Beach, and Oakland accounted for more than 40 percent of U.S. container shipping by volume (in twenty-foot equivalent units, or TEUs) and more than 14 percent ($370 billion) of the total foreign trade through the nation's top fifty international freight gateways. Data are from Bureau of Transportation Statistics, "State Transportation Statistics 2009," www.bts.gov/publications/state_transportation_statistics/.

3. Robert Wilkinson, "Preparing for a Changing Climate: The

Potential Consequences of Climate Variability and Change for California," Report for the U.S. Global Change Research Program, 2002.

4. Timmerman et al. argue that ENSO events will increase in frequency as temperatures warm. Cayan et al. find little change in the frequency of ENSO events. See Axel Timmermann et al., "Increased El Niño Frequency in a Climate Model Forced by Future Greenhouse Warming," *Nature* 398 (1999): 694–97; Daniel R. Cayan et al., "Climate Change Scenarios for the California Region," *Climatic Change* 87 (2008): S21–S42.

5. Landslides form a large share of storm costs. Porter et al. report that in a typical year Caltrans spends $20 million to $40 million repairing damage from landslides, and in wet years repair costs increase to around $150 million. In preparation for the 1997–98 El Niño–Southern Oscillation, Caltrans spent more than $200 million, most of which went toward fixing storm damage from the previous winter. See Keith Porter et al., "Overview of the ARkStorm Scenario," U.S. Department of the Interior and U.S. Geological Survey Open File Report 2010-1312, 2011; California Natural Resources Agency, "California El Niño Information, Lead El Niño Agencies in California," http://ceres.ca.gov/elnino/arar/bth.html.

6. Porter et al., "Overview of the ARkStorm Scenario," 18.

7. For instance, the Bureau of Transportation Statistics finds that California's roads are significantly poorer quality than the U.S. average, and that nearly 30 percent of its road bridges were either structurally deficient or functionally obsolete in 2009. See Bureau of Transportation Statistics, "State Transportation Statistics 2009."

8. American Society of Civil Engineers, "California Infrastructure Report Card 2006," www.ascecareportcard.org/.

9. California's 2009–10 budget allocates $15.7 billion for transportation spending, including all categories of line items 26 and 27, out of a total budget of $119.2 billion. See California Department of Finance, "California Budget 2009–10," http://2009-10.archives.ebudget.ca.gov/.

10. Michael J. Savonis, Virginia R. Burkett, and Joanne R. Potter, "Impacts of Climate Change and Variability on Transportation Systems and Infrastructure: Gulf Coast Study," Report by the U.S.

Climate Change Science Program and the Subcommittee on Global Change Research, 2008.

CHAPTER SIX

1. California Travel and Tourism Commission, "Facts and Figures," http://tourism.visitcalifornia.com/Research/.

2. See, for instance, Robert Mendelsohn and Marla Markowski, "The Impact of Climate Change on Outdoor Recreation," and John Loomis and John Crespi, "Estimated Effects of Climate Change on Selected Outdoor Recreation Activities in the United States," both in *The Impact of Climate Change on the United States Economy*, ed. Robert Mendelsohn and James Neumann (Cambridge: Cambridge University Press, 1999).

3. Stephen Saunders et al., "Losing Ground: Western National Parks Endangered by Climate Disruption," Rocky Mountain Climate Organization and Natural Resources Defense Council, New York, 2006.

4. Philip King, "Economic Analysis of Beach Spending and the Recreational Benefits of Beaches: San Clemente," Final Report to the City of San Clemente, 2001.

5. The 2000 National Survey on Recreation and the Environment (NSRE) estimates that 12.6 million Californians visited beaches in 1999–2000. See Vernon R. Leeworthy and Peter C. Wiley, "Current Participation Patterns in Marine Recreation," *National Survey on Recreation and the Environment 2000*, U.S. Department of Commerce, National Oceanic and Atmospheric Administration, and National Ocean Service, 2001. The Woods Hole Oceanographic Institution estimates that the direct revenue generated by beach recreation in California was $13 billion in 2001. King estimates that in 1998 direct revenue of California beaches was $14 billion and total economic benefits (direct, indirect, and induced) were $73 billion. Using a narrower population set and spending definition, Pendleton and Kildow estimate that California residents spend $4 billion a year on beach recreation. At the extreme, King and Symes estimate that the complete disappearance of beaches

would lead to a total annual economic loss of $8.3 billion to California. See Woods Hole Oceanographic Institution, "An Inventory of California Coastal Economic Sectors," National Oceanographic Partnership Program Report, 2003; Philip King, "The Fiscal Impact of Beaches in California," Final Report to the California Department of Boating and Waterways, 1999; Linwood Pendleton and Judith Kildow, "The Non-market Value of Beach Recreation in California," *Shore & Beach* 74 (2006): 34–37; Philip King and Douglas Symes, "The Potential Loss in Gross National Product and Gross State Product from a Failure to Maintain California's Beaches," Final Report to the California Department of Boating and Waterways, 2003.

6. This includes an average nine-meter reduction from a one-meter increase in sea levels, and an average ten-meter reduction from storm activity, at current sea level. See Linwood Pendleton et al., "Estimating the Potential Impacts of Climate Change on Southern California Beaches," California Climate Change Center Final Paper, CEC-500-2009-033-F, 2009.

7. In a worst-case scenario, assuming no increase in the beach-going population, Pendleton et al. estimate that a one-meter rise in sea levels would reduce annual attendance at beaches by 0.6 million, which, assuming at least thirteen million persons visit the beach annually, would be a reduction of less than 5 percent. This drop-off in tourism would lead to reduction in revenues of $15 million (in 2000 dollars), which, if total beach-related spending is at least $4 billion, would be less than 1 percent. For estimated impacts, see ibid.

8. Ibid., 26.

9. Ibid., 36.

10. This estimate is based on the average snow season from 1961 to 90, from Katharine Hayhoe et al., "Emissions Pathways, Climate Change, and Impacts on California," *Proceedings of the National Academy of Sciences* 101 (2004): 12422–27.

11. Hayhoe et al. provide the only projection of climate change impacts on the California ski industry. Under a B1 scenario using the PCM model, Hayhoe et al. estimate that the ski season would be delayed by twenty-two days and would be forty-nine days shorter by

the end of the century. Using the Hadley model with an A1fi emissions scenario, the ski season ceases to exist by the end of the century. Shaw et al. report IPCC projections of a reduction in ski season length from Field et al., but these projections are reportedly from Hayhoe et al. The estimates reported in Field et al. for 2050, moreover, do not match those in Hayhoe et al. Hayhoe et al. do not report detailed projections of reductions in the ski season for 2050. See Hayhoe et al., "Emissions Pathways," Supporting Text; M. Rebecca Shaw et al., "The Impact of Climate Change on California's Ecosystem Services," California Climate Change Center Final Paper, CEC-500-2009-025-F, 2009; C. B. Field et al., "North America. Climate Change 2007: Impacts, Adaptation and Vulnerability," in *Fourth Assessment Report of the Intergovernmental Panel on Climate Change,* ed. M. L. Parry et al. (Cambridge: Cambridge University Press 2007): 617–52.

12. The provenance of this estimate, though widely cited and used in PIER reports (e.g., Shaw et al.) is unclear. The California Ski & Snowboard Safety website reports that this estimate is based on a news release from the California Travel and Tourism Commission, but we were unable to locate an actual report to corroborate this estimate. See California Ski & Snowboard Safety, "Ski and Snowboard General Facts," www.calskisafety.org/reports/general-stats.html; Shaw et al., "The Impact of Climate Change," 48.

13. Changes in the ski season are likely to have a much larger impact on ski resort profits than on their revenues. Most of the expected shortening of the ski season results from earlier snowmelt, reducing the number of days at the tail of the season. However, a later onset of the ski season may be more difficult for ski resorts to overcome because of travel season and capacity considerations. Snowmaking would be one option for lessening impacts on revenues, but it would still reduce ski resorts' margins.

14. U.S. Fish and Wildlife Service, *2006 National Survey of Fishing, Hunting, and Wildlife-Associated Recreation: California* (Washington, DC: Fish and Wildlife Service, Department of the Interior, Department of Commerce, Economics and Statistics Administration, and Census Bureau, 2007).

15. National Parks Conservation Association, *National Treasures as Economic Engines* (Oakland: NPCA, 2003).

16. As an extreme example, in 1999 the federal government and state of California spent $480 million to purchase 7,472 acres of redwood forest, at more than $64,000 per acre, from Pacific Lumber. For more on this agreement, see Paul Rogers, "A Decade after Headwaters Deal, Truce Comes to Northern California Redwood Country," *Mercury News,* March 8, 2009.

17. U.S. Fish and Wildlife Service, *2006 National Survey.*

18. Dennis D. Lynch and John C. Risley, "Klamath River Basin Hydrologic Conditions Prior to the September 2002 Die-Off of Salmon and Steelhead," USGS Water-Resources Investigations Report 03-4099, 2003.

19. John G. Williams, "Central Valley Salmon: A Perspective on Chinook and Steelhead in the Central Valley of California," *San Francisco Estuary and Watershed Science* 4 (2006).

20. Daniel Scott, "The Impact of Climate Change on Golf Participation in the Greater Toronto Area (GTA): A Case Study," *Journal of Leisure Research* 38 (2006): 363–80.

21. Golf playable days are defined as those with a temperature of 45°F to 105°F and less than 0.25 inches of rainfall.

22. WeatherBill, "Impact of Climate Change on Golf Playable Days in the United States," WeatherBill Report, 2007.

23. Pendleton et al. estimate that the total cost of beach nourishment required to mitigate a one-meter rise in sea level would be $436 million, which, over the course of a century, works out to around $4 million per year. Kildow and Colgan report that local, state, and federal governments spent $10 million on beach nourishment in California during the 2000–2001 fiscal year. Pendleton et al. estimate that a stormy year would require $382 million in beach nourishment. See Pendleton et al., "Estimating the Potential Impacts," 39; Judith Kildow and Charles S. Colgan, "California's Ocean Economy," National Ocean Economic Program Report to the California Resources Agency, 2005.

24. We were unable to find published estimates of the cost of snowmaking in California. Gelt reports that the variable cost of snowmak-

ing in Arizona was $923 per acre-foot, which we assume is high relative to California. See Joe Gelt, "Got Snow? Effects of Climate Variability, Change on Arizona Skiing," *Arizona Water Resource* 14 (2006).

25. See note 16.

26. Pendleton et al. estimate that the loss in consumer surplus from a one-meter rise in sea levels and severe storms would be $40 million to $63 million and $37 million per year, respectively. By these estimates, and given the cost of beach nourishment, Pendleton et al. argue that nourishment is justified for dealing with inundation but probably not for severe storms. See Pendleton et al., "Estimating the Potential Impacts," 39, 43–44.

CHAPTER SEVEN

1. Data are from the Bureau of Economic Analysis (BEA) website, "Regional Data," http://bea.gov/iTable/iTable.cfm?reqid=70&step=1 &isuri=1&acrdn=3.

2. Two competing trends are at play. First, a warmer and wetter atmosphere increases storm intensity. Second, however, because the poles are warming faster than lower latitudes, the temperature difference between polar regions and midlatitude regions is decreasing, which would tend to weaken midlatitude storms. For modeled results see Daniel R. Cayan et al., "Climate Change Scenarios for the California Region," *Climatic Change* 87 (2008): S21–S42.

3. More specifically, in climate models the number of years with multiple major storms increases, though the total number of major storms does not change. See Michael Dettinger et al., "Projections of Potential Flood Regime Changes in California," California Climate Change Center Draft Paper, CEC-500-2009-050-D, 2009.

4. Ibid.

5. The California Department of Forestry and Fire Protection estimates that in California 11.8 million homes are in the wildland-urban interface (WUI), 4.9 million of which are exposed to high or greater fire threat. See California Board of Forestry, "Fuel Hazard Reduction Emergency Regulation," BOF Adopted Emergency Rule, 2004. This

document does not describe how the estimate of 4.9 million was calculated, and the estimate of 11.8 million homes in the WUI is not consistent with Radeloff et al.'s estimate of 5.1 million housing units in
the WUI. See V.C. Radeloff et al., "The Wildland-Urban Interface in
the United States," *Ecological Applications* 15 (2005): 799–805. The 40 percent estimate was calculated by dividing 4.9 million homes by California's 2006 housing stock of 13.1 million homes. Housing data are from
the California Department of Finance, "California County Profiles,"
www.dof.ca.gov/HTML/FS_DATA/profiles/pf_home.php.

6. Market value is the value a home would sell for on the open
market at current prices. The average market value here is based on
a statewide average value of $361,348 (in 2008 dollars) in 2010 and the
California Department of Forestry and Fire Protection's (CDF's) estimate of 4.9 million homes with high or greater fire threat. Housing
value data are from RAND California Business and Economic Statistics, "Housing Sales Prices and Transactions in California Counties
and Cities," http://ca.rand.org/stats/economics/houseprice.html.

7. Replacement cost is the cost of rebuilding the home, ignoring
land values. Average replacement costs here are based on a nineteen-
county study of earthquake risk in Northern California conducted by
Kircher et al., where the average replacement cost was estimated to be
$136.21 per square foot. See Charles A. Kircher et al., "When the Big
One Strikes Again: Estimated Losses Due to a Repeat of the 1906 San
Francisco Earthquake," *Earthquake Spectra* 22 (2006): S297–S339. From
2002 to 2010, the average area of California homes was 1,532 square feet.
Square footage data are from RAND California Business and Eco-
nomic Statistics, "Housing Sales Prices and Transactions."

8. This estimate is from Westerling and Bryant, who cite a 1995
California Fire Plan. We were unable to find more recent estimates
of fire suppression costs. See A.L. Westerling and B.P. Bryant, "Climate Change and Wildfire in California," *Climatic Change* 87 (2008):
S231–S249.

9. The lower range of estimates is for a B1 emissions scenario; the
higher range is for an A2 scenario. Large fires are defined as fires that
burn an area greater than two hundred hectares. See A.L. Westerling

et al., "Climate Change, Growth, and California Wildfire," California Climate Change Center Draft Paper, CEC-500-2009-046-D, 2009.

10. Westerling and Bryant, "Climate Change and Wildfire."

11. Benjamin Bryant and Anthony Westerling, "Potential Effects of Climate Change on Residential Wildfire Risk in California," California Climate Change Center Final Paper, CEC-500-2009-048-F, 2009.

12. Westerling and Bryant, "Climate Change and Wildfire."

13. Daniel R. Cayan et al., "Climate Change Projections of Sea Level Extremes along the California Coast," *Climatic Change* 87 (2008): S57–S73.

14. Dan Cayan et al., "Climate Change Scenarios and Sea Level Rise Estimates for the California 2009 Climate Change Scenarios Assessment," California Climate Change Center Final Paper, CEC-500-2009-014-F, 2009.

15. E. R. Thieler and E. S. Hammar-Klose, "National Assessment of Coastal Vulnerability to Future Sea-Level Rise: Preliminary Results for the U.S. Pacific Coast," U.S. Geological Survey, Open-File Report 00-178, 2000.

16. E. R. Thieler and E. S. Hammar-Klose, "National Assessment of Coastal Vulnerability to Future Sea-Level Rise: Preliminary Results for the U.S. Gulf of Mexico Coast," U.S. Geological Survey, Open-File Report 00-179, 2000.

17. James Neumann and N. D. Livesay, "Coastal Structures: Dynamic Economic Modeling," in *Global Warming and the American Economy: A Regional Analysis,* ed. Robert Mendelsohn (Cheltenham, UK: Edward Elgar, 2001), 142.

18. Based on the 2000 U.S. Census, 14 percent of the state's population lived and 16 percent of its housing stock was located "near shore." Data are from the National Ocean Economics Program (NOEP) website, www.oceaneconomics.org. NOEP defines "near shore" as "located in a zip code that is immediately adjacent to an ocean, Great Lake, or included river or bay."

19. The average market value here is based on statewide average value of $361,348 (in 2008 dollars) in 2010, and NOEP's estimate of three million homes near shore. Given that the value of near-shore houses

tends to be higher than average, $1 trillion is likely an underestimate but of the correct order of magnitude. Housing value data are from RAND California Business and Economic Statistics, "Housing Sales Prices and Transactions."

20. James E. Neumann et al., "Market Impacts of Sea Level Rise on California Coasts," in "Global Climate Change and California: Potential Implications for Ecosystems, Health, and the Economy," ed. Tom Wilson et al., Final Report to the Public Interest Energy Research Program, 500-03-058CF, 2003.

21. P.H. Gleick and E.P. Maurer, "Assessing the Costs of Adapting to Sea-Level Rise: A Case Study of San Francisco Bay," Pacific Institute for Studies in Development, Environment, and Security, 1990.

22. Matthew Heberger et al., "The Impacts of Sea-level Rise on the California Coast," California Climate Change Center Final Paper, CEC-500-2009-024-F, 2009.

23. National Climatic Data Center (NCDC), National Oceanic and Atmospheric Administration (NOAA), "1980–2010 Billion Dollar U.S. Weather Disasters," www.ncdc.noaa.gov/img/reports/billion/disasters2010.pdf.

24. Keith Porter et al., "Overview of the ARkStorm Scenario," U.S. Department of the Interior and U.S. Geological Survey Open File Report 2010-1312, 2011.

25. More specifically, the authors estimate that such a storm would result in more than $300 billion in direct property damages ($200 billion building, $100 billion contents), $60 billion in "demand surge" as repair costs increase, $40 billion in agricultural losses, damage to lifeline services, and other repair costs, and $325 billion in business interruption costs. Earthquake costs are based on the USGS ShakeOut Scenario studies. See ibid., v.

26. Porter et al. report that analysis of sediment in the San Francisco Bay and Santa Barbara indicates that large storms occurred in the years 212, 440, 603, 1029, 1418, and 1605 CE, but they do not provide a reference. Over two millennia, this frequency implies an average return period of roughly three hundred years. See ibid., 2.

27. Howard C. Kunreuther and Erwann O. Michel-Kerjan, "Cli-

mate Change, Insurability of Large-scale Disasters and the Emerging Liability Challenge," National Bureau of Economic Research Working Paper 12821, 2007.

28. For a succinct history of the evolution of catastrophic risk in the insurance industry, see Michael Lewis, "In Nature's Casino," *New York Times,* August 26, 2007.

29. Kunreuther and Michel-Kerjan, "Climate Change."

30. Marc Lifsher, "California Homeowners Facing Insurance Rate Hikes," *Los Angeles Times,* September 1, 2009.

31. Established by the California legislature in the wake of the Watts riots, FAIR was originally intended to provide affordable coverage in urban neighborhoods deemed too risky to insure.

32. Information on the extent of coverage under FAIR is not available from the FAIR website. A *Los Angeles Times* report indicates that more than 180,000 of the 6.5 million owner-occupied homes in the state (3 percent) were covered under FAIR in 2003, though this number has likely gone up. In 2003 more than 80 percent of policyholders were reportedly in Southern California, including high-end neighborhoods such as Malibu and Bel Air, which has made the plan controversial. See Jeff Bertolucci, "Hard to Insure," *Los Angeles Times,* September 28, 2003.

33. Ellen Hanak et al., *Managing California's Water: From Conflict to Reconciliation* (San Francisco: Public Policy Institute of California, 2011), 103–5.

34. Porter et al., "Overview of the ARkStorm Scenario," 100–101.

35. Erwann Michel-Kerjan and Howard Kunreuther, "Redesigning Flood Insurance," *Science* 333 (2011): 408–9.

36. J. David Cummins, Michael Suher, and George Zanjani, "Federal Financial Exposure to Natural Catastrophe Risk," in *Measuring and Managing Federal Financial Risk,* ed. Deborah Lucas (Washington, DC: University of Chicago Press, 2010).

37. Max A. Moritz and Scott L. Stephens, "Fire and Sustainability: Considerations for California's Altered Climate." *Climatic Change* 87 (2008): S265–S271.

38. Hanak et al., *Managing California's Water,* 290–306.

39. Heberger et al., "The Impacts of Sea-level Rise," xi.

CHAPTER EIGHT

1. The Lancet, "A Commission on Climate Change," *The Lancet* 373 (2009): 1659.

2. According to the American Lung Association, eight of the ten cities with the worst ozone are in California, as are six of the ten with the worst year-round particle pollution. See American Lung Association, *State of the Air 2011,* www.stateoftheair.org/2011/city-rankings/most-polluted-cities.html.

3. California Air Resources Board and American Lung Association of California, "Recent Research Findings: Health Effects of Particulate Matter and Ozone Air Pollution," CARB and ALAC Policy Brief, 2004.

4. Deborah Drechsler et al., "Public Health–Related Impacts of Climate Change in California," California Climate Change Center White Paper, CEC-500-2005-197-SF, 2006.

5. Ozone (O_3) forms from the reaction of diatomic oxygen (O_2) with an oxygen molecule (O). The main source of O for this reaction is nitrogen dioxide (NO_2), which disassociates into NO_2 + sunlight. Power plants and vehicles are a large source of nitric oxide (NO), but a relatively small amount of this NO is directly oxidized to NO_2. During their decomposition in the atmosphere, VOCs react with this NO to create NO_2, the "fuel" for ozone. The main removal mechanism for ozone is NO + O_3 → NO_2 + O_2, which hints at the difficulty of ozone control. Reducing NO_x emissions, for instance, can increase or reduce ozone concentrations, depending on the concentrations of NO_2 and NO.

6. Michelle L. Bell et al., "Ozone and Short-term Mortality in 95 US Urban Communities, 1987–2000," *Journal of the American Medical Association* 292 (2004): 2372–78.

7. Rob McConnell et al., "Asthma in Exercising Children Exposed to Ozone: A Cohort Study," *Lancet* 359 (2002): 386–91.

8. Environmental Protection Agency, "The Green Book Nonattainment Areas for Criteria Pollutants," www.epa.gov/airquality/greenbook/index.html.

9. Ozone formation in California tends to be NO_x limited, which means that although VOCs are abundant, NO_x is in relatively short

supply. Thus increasing VOC emissions will not necessarily lead to higher ozone levels without higher NO_x emissions.

10. Water vapor affects ozone chemistry through the supply of hydroxyl radicals (•OH), which form through the reaction of water (H_2O) with an oxygen (O) molecule, or $H_2O + O \rightarrow$ •OH + •OH. Hydroxyl radicals attack VOCs and initiate their decomposition, which leads to NO_2 creation. Changes in atmospheric circulation affect ozone concentrations primarily through the frequency and severity of atmospheric stagnation events, where limited air mixing occurs.

11. Drechsler et al. report that background mean ozone concentrations in California range from 15 to 50 parts per billion (ppb), and that the IPCC estimates that global background ozone concentrations could reach 40 to 80 ppb by 2100. California's eight-hour ozone standard is currently 70 ppb. See Drechsler et al., "Public Health–Related Impacts"; and California Air Resources Board, "Ambient Air Quality Standards," www.arb.ca.gov/research/aaqs/aaqs2.pdf.

12. The California Air Resources Board reports that ozone causes eight hundred premature deaths per year. Assuming, for the sake of illustration, that the relationship between ozone concentrations and mortality is linear, a 10–20 percent increase in ozone would lead to 80–160 premature deaths per year. At the EPA's central statistical value of life (SVL) estimate of $7.4 million (in 2006 dollars), this increase in mortality would be equivalent to societal costs of $0.6 billion to $1.1 billion per year. Hospitalization costs would rise, but, because ozone levels are not expected to be higher than present levels and ozone accounts for only part of air pollution health impacts, they would not exceed current total costs of around $2 billion per year. For the EPA's statistical value of life estimates, see EPA, National Center for Environmental Economics, "Frequently Asked Questions on Mortality Risk Valuation," http://yosemite.epa.gov/ee/epa/eed.nsf/pages/Mortality RiskValuation.html.

13. Rupa Basu, Feng Wen-ying, and Bart D. Ostro, "Characterizing Temperature and Mortality in Nine California Counties," *Epidemiology* 19 (2008): 138–45.

14. Rupa Basu and Paul English, "Public Health Impacts from Climate Change," Draft CEC PIER-EA Discussion Paper, 2008.

15. Ibid.

16. Drechsler et al., "Public Health–Related Impacts."

17. Gerald A. Meehl and Claudia Tebaldi, "More Intense, More Frequent, and Longer Lasting Heat Waves in the 21st Century," *Science* 305 (2004): 994–97.

18. More specifically, Drechsler et al. estimate that the likelihood of extreme heat would increase by a factor of 2 to 2.5 under a B1 emissions scenario, and by a factor of 2.5 to 4 under A2 and A1fi emission scenarios. See Drechsler et al., "Public Health–Related Impacts."

19. Drechsler et al. project increases in heat-related mortality in Fresno, Los Angeles, Sacramento, San Bernardino, and San Francisco ranging from 67 persons per year in an acclimated PCM B1 scenario to 837 persons per year in an unacclimated PCM A2 scenario in the 2035–64 period, and 183 persons per year in a PCM B1 scenario to 1,072 persons per year in a Hadley A1fi scenario in the 2070–99 period. These five counties accounted for 41 percent of California's population in 2000. Assuming that the average mortality rate, in deaths per million persons, is the same between the weighted average of these five counties and the state, state mortality is the mortality from these five counties divided by their share of the population. With this assumption, state mortality would range from 160 to 2,000 persons in 2035–64 and from 780 to 2,600 persons per year in 2070–99. At the $7.4 million SVL used previously, this increase in mortality would lead to social costs ranging from $1 billion to $15 billion in 2035–64 and $3 billion to $19 billion in 2070–99. See Drechsler et al., "Public Health–Related Impacts." County population data are from the California Department of Finance, "California County Profiles," www.dof.ca.gov/HTML/FS_DATA/profiles/pf_home.php.

20. Drechsler et al., "Public Health–Related Impacts," 61.

21. Basu and English, "Public Health Impacts from Climate Change."

22. Julian D. Marshall, "Environmental Inequality: Air Pollution Exposures in California's South Coast Air Basin," *Atmospheric Environment* 42 (2008): 5499–5503.

23. Bob McCormick, "Effects of Biodiesel on NO_x Emissions," Presentation at CARB Biodiesel Workshop, June 8, 2005.

INDEX